THE
PICTORIAL HISTORY
OF
WORLD
SPACECRAFT

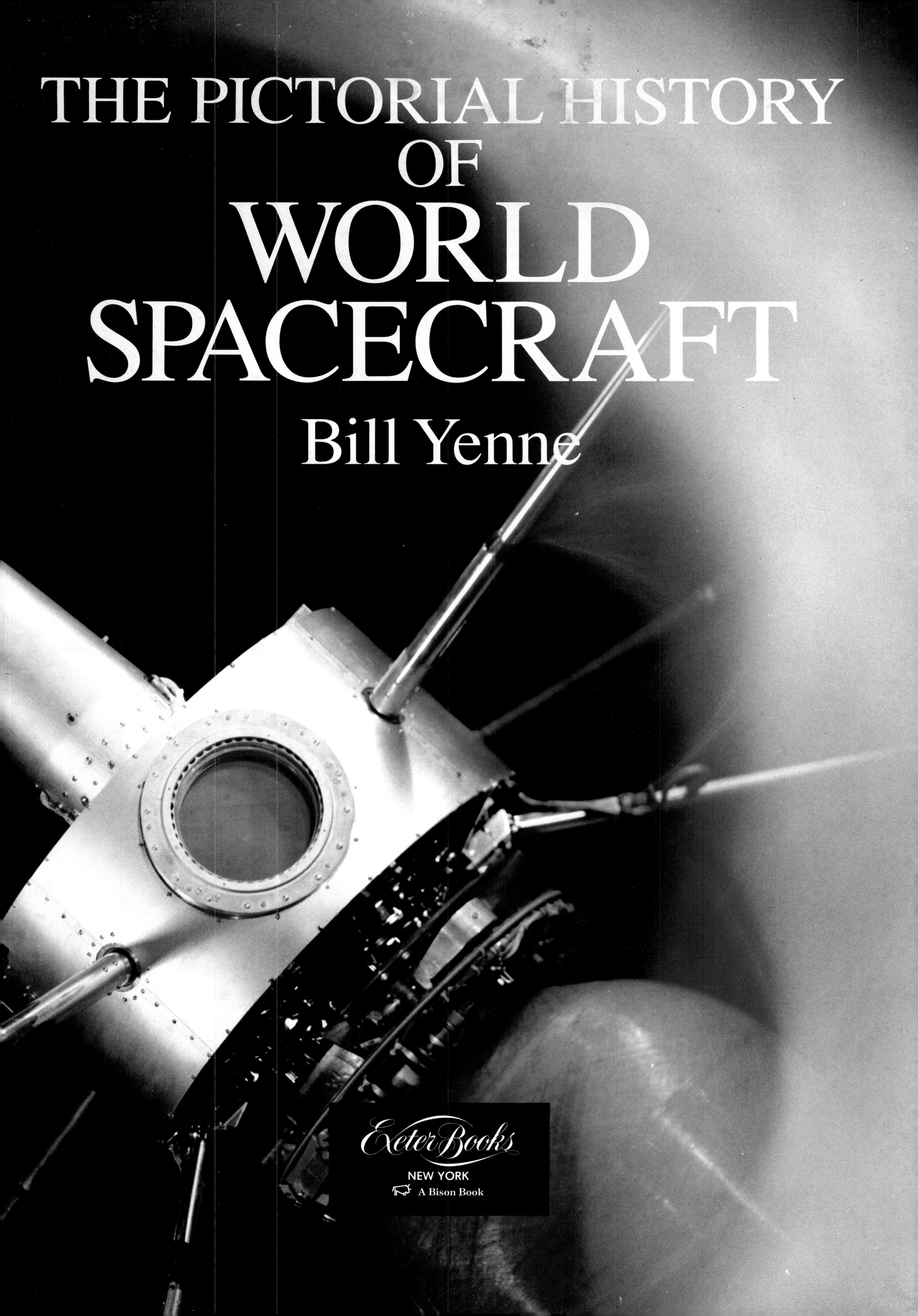

THE PICTORIAL HISTORY
OF
WORLD
SPACECRAFT

Bill Yenne

Exeter Books

NEW YORK

A Bison Book

Photo Credits

All photos were supplied through the generosity of the National
Aeronautics and Space Administration with the following excep-
tions:

Boeing Aerospace Company 58 (right), 69 (bottom center and
  top), 81 (above right), 112 (above)
Centre National d'Etudes Spatiales 155 (above), 184 (below)
Department of Defense 126, 150-151 (left), 189, 200 (above)
European Space Agency 194
© RF Gibbons 53 (all), 75 (above), 78-79 (below), 106, 125,
  155 (below), 196 (both)
Hughes Aircraft Company 26, 30-31 (all), 34 (above, both), 35,
  43, 44 (right center), 80, 90, 91 (above and below left), 95,
  97, 100-101 (all), 110, 113, 119 (above and below left), 121,
  133, 150-151 (right), 164, 165 (above), 173 (top and bottom
  left, right), 175 (below), 187 (below)
Lockheed Corporation 39 (above)
Max Planck Institut 197
McDonnell Douglas Corporation 15 (below), 18 (above), 22,
  33, 36-37 (both), 74
National Oceanic and Atmospheric Administration 122-123
National Space Development Agency of Japan 177
Novosti 8, 12 (right), 13
Rockwell International Corporation 68, 112 (below), 124
  (below), 141, 144, 145 (above), 153 (right) 170, 174, 175

(above, both), 176 (above right), 182, 183, 190 (lower right),
  191 (center), 208
Smithsonian Institution National Air & Space Museum 10
  (below), 12 (left), 44 (below, both), 45, 57, 58 (left), 60-61
  (both), 72 (below), 146-147, 156 (top)
TASS from Sovfoto 19, 24, 28, 115, 124 (above), 131 (above),
  156-157 (except 156 top), 184-185 (main photo)
© DR Woods 56 (both), 77 (above), 93 (below right)
© Bill Yenne 69 (right), 130, 131 (below), 185 (below)

Designed by Bill Yenne
**Edited by Pamela Berkman (with our thanks to RF
  Gibbons for reviewing the manuscript)**
Captioned by Timothy Jacobs

Page 1: Pioneer 11 and a mockup of its launching shroud is
shown during an inspection previous to its 6 April 1973 launch.
Pioneer 11 sent back detailed information on Saturn and Jupiter,
and with its sister craft Pioneer 10, set the stage for the spectacular
Voyager spacecraft discoveries.

Pages 2-3: Explorer 20, launched on 25 August 1964, made radio
soundings of Earth's upper atmosphere and collected data on atmospheric ionization irregularities. The craft is shown here with
its antenna arms extended, as it would appear in orbit. Note the
experiment sphere atop Explorer 20's truncated cone.

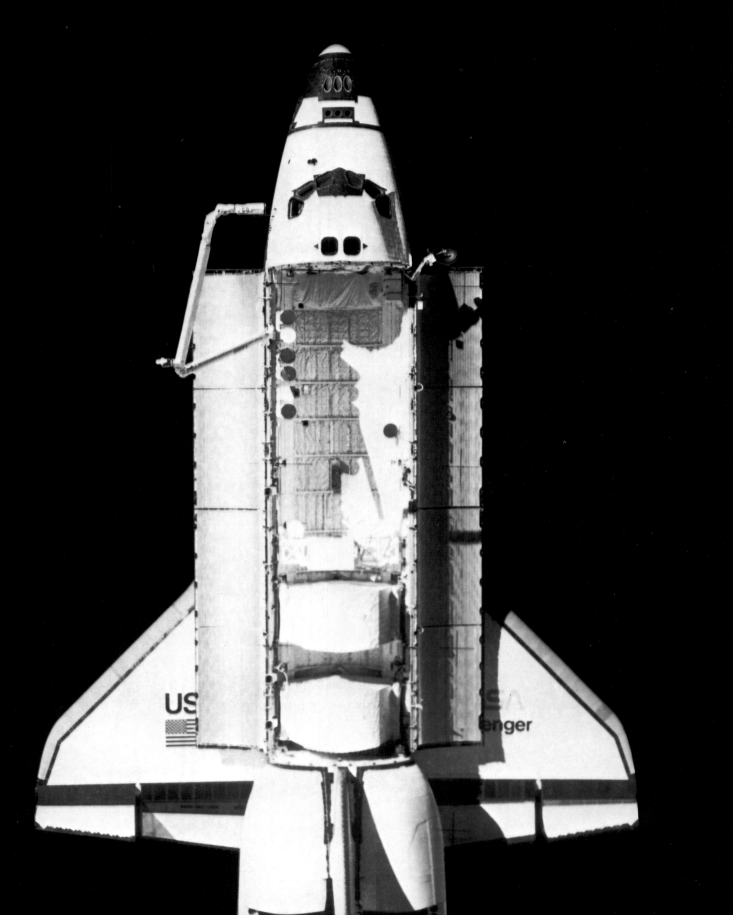

TABLE OF CONTENTS

PART ONE

The Formative Years: 1957–1960

7

PART TWO

The Dawn of Manned Space Flight: 1961–1966

17

PART THREE

The Race to the Moon: 1967–1972

41

PART FOUR

The Age of the Space Station: 1973–1980

71

PART FIVE

The Age of the Space Shuttle: 1981–1984

139

PART SIX

Space Flight Comes of Age: 1985–1987

191

Part One

The Formative Years

1957–1960

1957

It was the year that space flight became a reality. On 4 October, a be-whiskered sphere called Sputnik 1 was launched into earth orbit from the Soviet launch facility at Tyuratam. Weighing 184 pounds, the gleaming metallic ball carried instrumentation designed to measure the density of the atmosphere and had a radio transmitter that allowed it to be tracked in its orbit, which had an average altitude of 300 miles. The spacecraft continued to transmit for three weeks and remained in orbit for 96 days, finally reentering the earth's atmosphere after 1400 orbits.

This much heralded Soviet success was followed less than a month later by the launch of Sputnik 2 on 3 November. Within a special pressurized sphere, this second spacecraft carried the dog Laika, who had the distinction of being the first living creature to survive in space. Remote sensors monitored Laika's physiological reaction to the rigors of space travel and weightlessness until the oxygen in the sphere ran out seven days later and the poor creature expired. As for Sputnik 2, it remained in orbit for 103 days and 2370 orbits.

In the United States, meanwhile, efforts to place an artificial satellite into orbit failed during 1957. The American space program was the victim of a parochial interservice rivalry that saw the US Navy's Vanguard launch rocket pitted against the US Army's Redstone and the US Air Force's Atlas. The latter was sidetracked toward development as an ICBM and the Vanguard was chosen as first priority over the Redstone. Despite promising early results, the Vanguard was plagued by a series of disastrous failures through the fall of 1957 and the United States ended the year without placing a spacecraft into orbit.

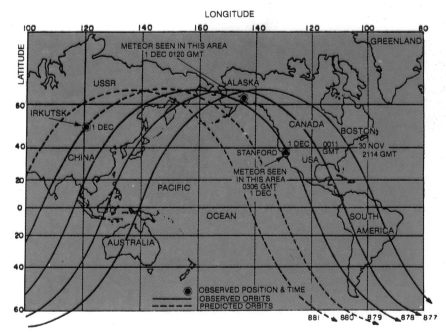

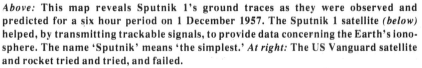

Above: This map reveals Sputnik 1's ground traces as they were observed and predicted for a six hour period on 1 December 1957. The Sputnik 1 satellite *(below)* helped, by transmitting trackable signals, to provide data concerning the Earth's ionosphere. The name 'Sputnik' means 'the simplest.' *At right:* The US Vanguard satellite and rocket tried and tried, and failed.

Space Flight Highlights of 1957

	Launch date	Launch vehicle	Launch weight
Sputnik 1 (USSR)	4 Oct	A	184 lb
Sputnik 2 (USSR)	3 Nov	A	1118 lb

IV

1958

After beginning the year 2-0 behind the Soviet Union in what was now dubbed the 'Space Race', the United States made a remarkable comeback by switching from the US Navy Vanguard launch vehicle to the US Army Jupiter C, a launch vehicle derived from the Redstone rocket. On 1 February, a Jupiter C placed America's first satellite—Explorer 1—into orbit from Cape Canaveral, Florida. Despite its having come on the heels of two Soviet competitors, Explorer 1 achieved much more from a scientific and technological standpoint than the Sputniks. Not only did the Yank spacecraft continue to transmit data until 23 May, it proved stable enough to remain in orbit for *twelve years*. Its most significant achievement, however, was its confirmation of the existence of the earth's Van Allen radiation belt.

Explorer 1 was followed by an unsuccessful Explorer 2 launch on 5 March, and by a successful Navy Vanguard launch on 17 March. Dubbed Vanguard 1 despite having been preceded by a series of Vanguard failures, this tiny spacecraft continued to transmit data for seven years. Having emerged the winner in the interservice contest in the United States, the Army went on to successfully launch Explorer 4 on 26 July and Explorer 6 on 7 August.

By this time the United States wisely chose to end the contest between Navy Vanguard and Army Explorer by channeling the American space flight effort via the National Aeronautics and Space Administration (NASA) which was formally established on 1 October. Less than two weeks later, on 11 October, NASA launched its first spacecraft, the former US Air Force Pioneer 1. Designed to reach the moon, Pioneer 1 failed in this attempt but it did reach 70,000 miles into space (a third of the distance to the moon) before returning to earth.

The Soviet Union succeeded with only a single satellite during 1958. Sputnik 3 was launched on 15 May and remained in orbit for 690 days.

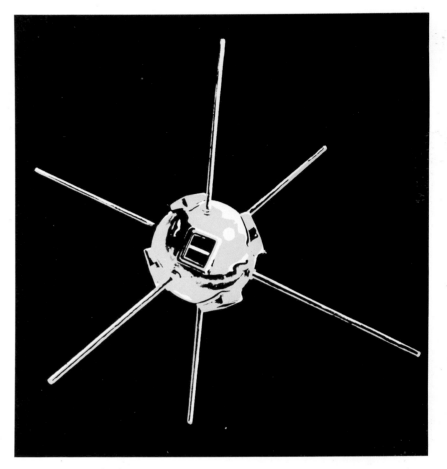

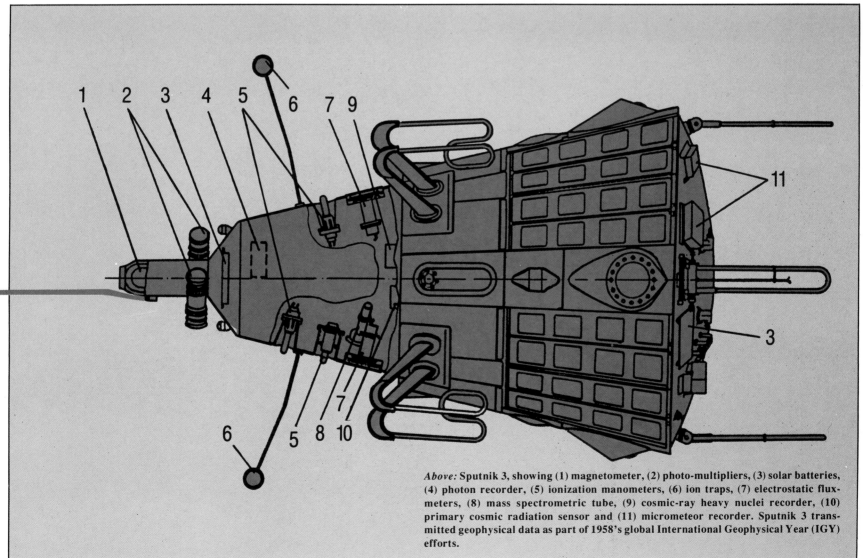

Above: Sputnik 3, showing (1) magnetometer, (2) photo-multipliers, (3) solar batteries, (4) photon recorder, (5) ionization manometers, (6) ion traps, (7) electrostatic flux-meters, (8) mass spectrometric tube, (9) cosmic-ray heavy nuclei recorder, (10) primary cosmic radiation sensor and (11) micrometeor recorder. Sputnik 3 transmitted geophysical data as part of 1958's global International Geophysical Year (IGY) efforts.

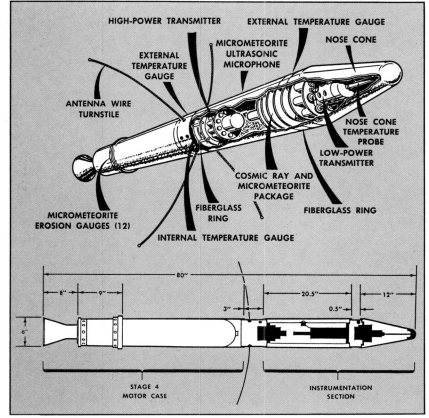

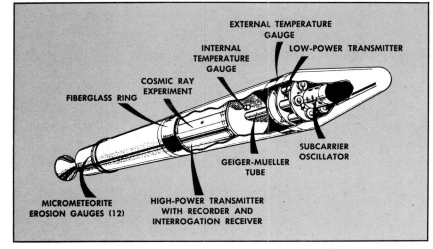

Above left: Success at last for Vanguard! Vanguard 1—not the first, but the first successful Vanguard vehicle launch—not only made it into orbit, but transmitted data for seven years. *At right:* The Explorer 1, launched atop a US Army Jupiter C booster, confirmed the existence of the Earth's Van Allen radiation belt, located at an altitude of 600 miles. *At top and above:* Cutaway views of, respectively, Explorer 1 and Explorer 3—the former *(top)* was the first American satellite to achieve orbit.

Space Flight Highlights of 1958

	Launch date	Launch vehicle	Launch weight
Explorer 1 (USA)	1 Feb	Jupiter C	31 lb
Vanguard 1 (USA)	17 Mar	Vanguard TV4	3 lb
Sputnik 3 (USSR)	15 May	Vostok	2919 lb
Pioneer 1 (USA)	11 Oct	Thor-Able	75 lb
Pioneer 3 (USA)	6 Dec	Juno 2	13 lb

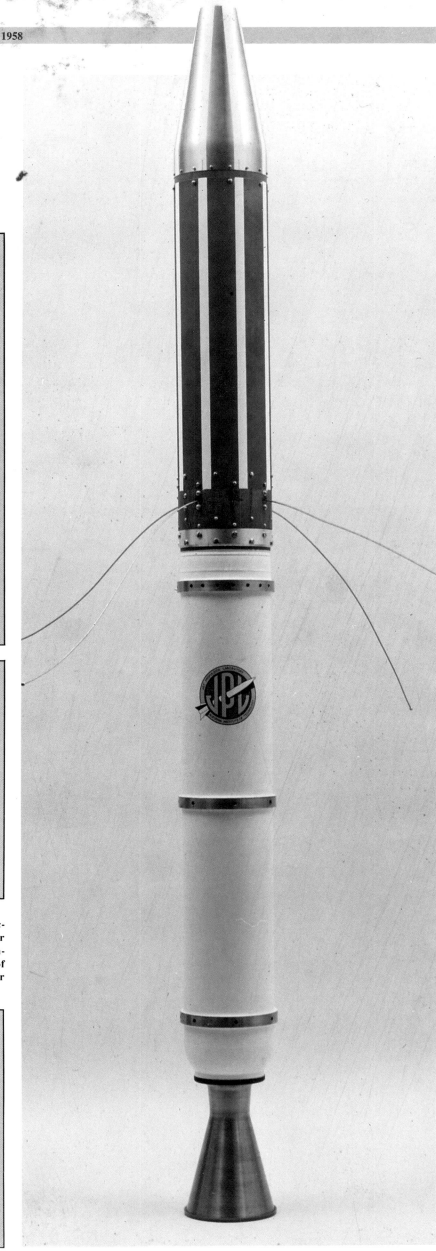

1959

The Soviet Union began the third year of the space race on a high note on 2 January with its Luna 1 which came to within 3100 miles of the moon. If 1957 was the year of the Sputniks and 1958 the year of the Explorers, then 1959 was certainly the year of the Soviet Union's Luna series spacecraft. Luna 1 was followed by Luna 2 which was launched on 12 September and became the first man-made object to reach the moon when it crashed into the Sea of Serenity a few days later. Less than a month after this—on 4 October—Luna 3 set out on the trail of its older brothers in what was to be the most complex trajectory yet accomplished by a spacecraft. By placing Luna 3 into an elliptical orbit around the earth that had an apogee of nearly 300,000 miles, the Soviet Union was able to actually include the moon *within* that orbit. As a result, Luna 3 slipped behind the moon. For countless centuries, mankind had stared up at the visible near side of the moon, wondering what might lie on the far side. In October 1959, the television cameras of Luna 3 transmitted the first fuzzy photos of that mysterious hemisphere.

For its part, the United States launched two successful Vanguards in February and September, and two successful Explorers in August and October. The Pioneer, which was successfully launched on 3 March, only came to within 37,300 miles of the moon, a pale achievement by comparison to the success of the Lunas.

By now, NASA had embarked on its heavily publicized Mercury Project, designed to put a man into space. Project activities during 1959 included a series of test launches of empty Mercury spacecraft and a successful suborbital launch of a space-rated Mercury capsule carrying a monkey named Sam.

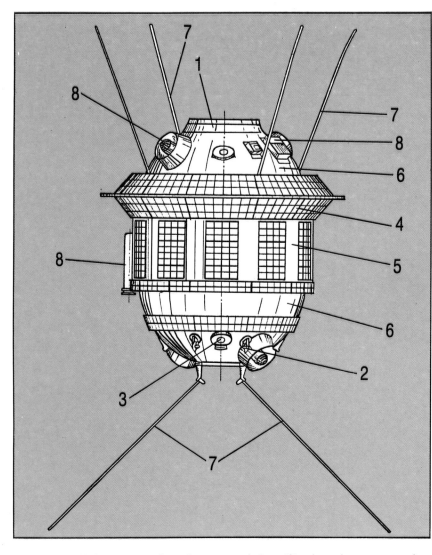

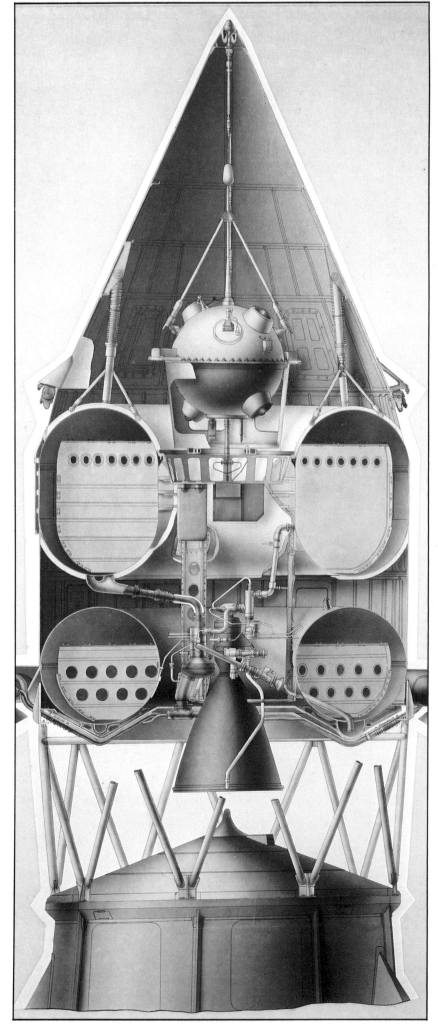

Above: Luna 3's instrumentation: (1) camera window, (2) orientation system engine, (3) solar sensor, (4) solar battery section, (5) shutters of temperature control system, (6) heat shields, (7) aerials and (8) scientific instruments. *At right:* Luna 1 in its protective launch cone. *Opposite:* Models of Luna 1 (near) and Luna 3 on display in Moscow, at the 'Cosmonautics Memorial Museum.'

Space Flight Highlights of 1959

	Launch date	Launch vehicle	Launch weight
Luna 1 (USSR)	2 Jan	A-1	794 lb
Luna 2 (USSR)	12 Sep	A-1	858 lb
Luna 3 (USSR)	4 Oct	A-1	957 lb
Mercury/Little Joe 3 (USA)	4 Dec	Little Joe	2000 lb

1960

After three years of what might be called 'experimental' spacecraft, the new decade ushered in the era in which unmanned space flight could be said to have become routine.

On 1 April, the United States launched TIROS 1, the world's first weather satellite. It operated for a little over two months, obeying commands from its controllers on earth and sending back television pictures of cloud formations and storm systems.

The United States followed this 'first' with the first navigation satellite (Transit 1B) on 13 April, the first infrared surveillance satellite (Midas 2) on 24 May, and the first passive communications satellite (Echo 1) on 12 August. Echo 1 was a 100 foot inflatable, aluminized mylar sphere that successfully reflected radio waves from earth-based transmitters in the United States to receivers in England.

For the Soviet Union, 1960 was a mixed bag of success and failure as it grasped for new plateaus in space exploration. On 19 August, Sputnik 5 went into space carrying the dogs Belka and Strelka. The following day, after 18 orbits, the two canines became the first living creatures to be recovered successfully from an orbital space flight. On 1 December, Sputnik 6 with the dogs Mushka and Pchelka was launched in hopes of repeating the success of the August launch, but it was not to be. A malfunction in the spacecraft's retrorocket system resulted in too steep an angle of reentry and the animals perished when Sputnik 6 burned up.

Below: **A view of TIROS 1, the world's first weather satellite, launched with a Thor booster. TIROS 1's two television cameras relayed 22,952 photos of cloud cover and other valuable meteorological phenomena between 1 April and 29 June, when the satellite ceased transmission to Earth sensors.**

Space Flight Highlights of 1960

	Launch date	Launch vehicle	Launch weight
TIROS 1 (USA)	1 Apr	Thor	270 lb
Echo 1 (USA)	12 Aug	Thor-Delta	166 lb
Sputnik 5 (USSR)	19 Aug	Vostok	10,120 lb

At top of page:—more than *just* a 100 ft mylar balloon— Echo 1 was the world's first passive communications satellite, and helped transmit two-way voice and television signals. *At right:* These Mercury capsules, in process on McDonnell's St Louis production line, portended a future of manned space flight.

Part Two

The Dawn of Manned Space Flight

1961–1966

1961

I n the fifth year of mankind's reach into space, man himself at last traveled into the chilly nether reaches beyond the thin blanket of earth's atmosphere.

The Soviet Union had launched and recovered dogs from prototypes of what would be its manned space capsules, and the United States had done the same on three suborbital test flights with monkeys. The last such US flight took place on 31 January with the chimpanzee Ham. The Soviet dogs Chernoshka and Zvesdochka traveled into space on single-orbit flights on 9 and 25 March.

By now both countries, nervous that the other would be first with a man in space, were ready to stop monkeying around and get on with the space missions that had fired everyone's imagination.

On 12 April Soviet test pilot Yuri Gagarin became his country's first cosmonaut and the first human being in space. Flying aboard Vostok 1, a variation of the later Sputniks of 1960, Gagarin completed a single orbit and returned to a hero's welcome.

Less than a month later on 5 May, Alan Shepard became the first American 'astronaut' in space. Launched into space by a Redstone launch vehicle, Shepard flew aboard the Mercury 3 spacecraft which he had dubbed *Freedom 7*. The '7' was a reference to the number of men in the first American astronaut corps. All of the Mercury capsules would be given nicknames ending in that numeral.

The second American manned space flight took place on 21 July with Virgil 'Gus' Grissom at the controls of Mercury 4. Grissom's suborbital flight was followed by the second Soviet manned flight in which 17 orbits of earth were completed.

The subsequent orbital test flight of Mercury 5, with the chimpanzee Enos aboard, was less than successful when it had to be brought back to earth early due to an excessive rate of roll.

In the first year of manned space flight, the Soviet Union and the United States had each put two men into space but the Soviet cosmonauts had an aggregate total of 18 orbits, while neither of the American astronauts had flown even a single complete orbit.

Space Flight Highlights of 1961

	Launch date	Launch vehicle	Launch weight
Vostok 1 *(Swallow)* (USSR)	12 Apr	A-1	10,395 lb
Mercury 3 *(Freedom 7)* (USA)	5 May	Redstone	2000 lb
Mercury 4 *(Liberty Bell 7)* (USA)	21 Jul	Redstone	2000 lb
Vostok 2 *(Oriel)* (USSR)	6 Aug	A-1	10,408 lb

American Astronauts
1.	Alan Shepard	(Mercury 3)	5 May
2.	Virgil Grissom	(Mercury 4)	21 Jul

Soviet Cosmonauts
1.	Yuri Gagarin	(Vostok 1)	12 Apr
2.	German Titov	(Vostok 2)	6 Aug

At left: Trained to go into space aboard the Mercury series of spacecraft, the first American astronaut group included (front row, left to right) Walter M 'Wally' Schirra Jr, Donald K 'Deke' Slayton, John H Glenn Jr, M Scott Carpenter, (back row, left to right) Alan B Shepard Jr, Virgil 'Gus' Grissom and L Gordon Cooper. All but Slayton went into space during the Mercury program. *At top of page:* Mercury 3, including the 'Freedom 7' capsule and Alan Shepard, rises into the wild blue atop its Redstone rocket. *At right:* The first man in orbit, Yuri Gagarin, at home with his daughters. Gagarin died in a jet airplane accident in March 1968, while training for Soyuz 3.

1962

As in 1961, manned space flight received more attention than unmanned space flight, but as always, unmanned missions also played an important role. On the American side there was Telstar 1, the first commercial spacecraft. It was a communications satellite produced by American Telephone and Telegraph and launched by NASA on 10 July. Telstar was not only capable of transmitting telephone calls but, for the first time in history, live television pictures could be transmitted between the United States and Europe.

Both the United States and the Soviet Union launched their first attempts at planetary exploration in 1962. The American Mariner 2 was launched on 27 August (Mariner 1 had failed on 22 July) and the spacecraft flew within 21,648 miles of Venus on 14 December, transmitting a great deal of useful data. The Soviet Mars 1 probe, launched on 1 November, didn't fare as well. While it probably came close to the red planet, all communications with the spacecraft were lost four months after launch.

In the venue of manned space flight the United States launched Mercury 6 on 20 February, carrying John Glenn on three orbits of the earth. America's first orbital flight was followed by that of Scott Carpenter, who was launched on 24 May for three orbits, and that of Walter 'Wally' Shirra, who completed six orbits on 3 October.

The Soviet Union launched only two cosmonauts in 1962, but their flights continued the pattern of long duration missions that had begun in 1961 and added the new twist of having two men in space simultaneously. Andrian Nikolayev was launched on 11 August for 64 orbits, and he was joined the following day by Pavel Popovich, who completed 48 orbits, before both cosmonauts returned to earth independently on 15 August.

At left: The first American to orbit the Earth was Mercury 6 astronaut John Glenn. M Scott Carpenter became the second American in orbit—aboard Mercury 7, shown *above* during liftoff. *At right:* The first commercial satellite, AT&T's Telstar, which even inspired a pop music recording.

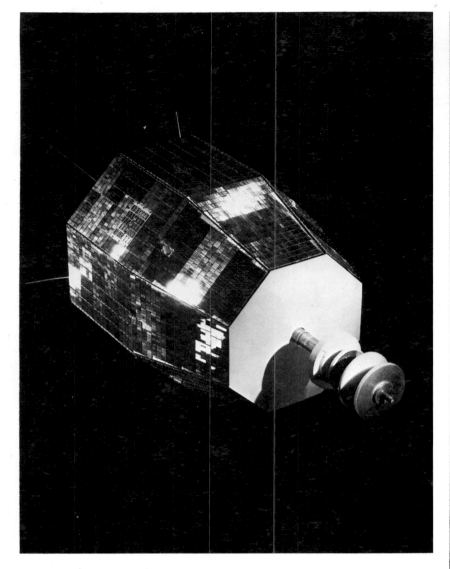

At left: America's first manned orbital spaceflight, Mercury 6 blasts off on 20 February 1962. The Relay 1 satellite is shown *above* at launching. Relay 1 relayed international television and voice microwave communications, and measured radiation levels, transmitting data until 2 February 1965. *At right:* The Ranger 4 Moon probe rises atop its Atlas-Agena, on 23 April 1962. Though its control systems failed, Ranger 4 became the first NASA payload to impact the Moon: it landed on the lunar far side, returning a few television images.

Space Flight Highlights of 1962

	Launch date	Launch vehicle	Launch weight
Mercury 6 *(Friendship 7)* (USA)	20 Feb	Atlas D	3000 lb
Ranger 4 (USA)	23 Apr	Atlas/Agena B	730 lb
Mercury 7 *(Aurora 7)* (USA)	24 May	Atlas D	3000 lb
Telstar 1 (USA)	10 Jul	Thor-Delta	170 lb
Vostok 3 *(Falcon)* (USSR)	11 Aug	A-1	10,388 lb
Vostok 4 *(Golden Eagle)* (USSR)	12 Aug	A-1	10,388 lb
Mercury 8 *(Sigma 7)* (USA)	3 Oct	Atlas D	3000 lb
Relay 1 (USA)	13 Dec	Thor-Delta	172 lb

American Astronauts
3. John Glenn (Mercury 6) 20 Feb
4. M Scott Carpenter (Mercury 7) 24 May
5. Walter Schirra (Mercury 8) 3 Oct

Soviet Cosmonauts
3. Andrian Nikolayev (Vostok 3) 11 Aug
4. Pavel Popovich (Vostok 4) 12 Aug

1963

The third year of manned space flight also saw an unheralded, routine continuation of earth-orbit unmanned scientific spacecraft. By the end of 1963, the Soviet's workhorse Cosmos series had progressed through Cosmos 23, while the similar American series saw the successful launch of Explorer 19 on 19 December. Meanwhile, a second Telstar was launched on 7 May, and on 14 February and 26 July, Syncom 1 and Syncom 2 were launched. The Syncoms were communications satellites, the first spacecraft ever placed into geosynchronous orbit—meaning that their orbits matched the rotation of the earth so that they 'hovered' over the same spot on the earth continuously.

During 1963, the Soviet Union and the United States also each concluded their first series of manned spacecraft flights.

Project Mercury ended with the 22 orbit flight of Gordon Cooper who went into space aboard Mercury 8 *(Faith 7)* on 15 May. A seventh planned Mercury flight was canceled in order to devote more of NASA's resources to the sub-sequent Project Gemini and because the seventh Mercury astronaut, Donald 'Deke' Slayton, developed a heart murmur.

The Soviet Vostok manned spacecraft series ended with another spectacular simultaneous flight of two capsules. Launched on 14 June aboard Vostok 5, cosmonaut Valery Bykovsky was followed two days later by Valentina Tereshkova, the first woman in space. The two cosmonauts completed 81 and 48 orbits respectively, returning to earth on 19 June.

At the end of the Mercury and Vostok programs the astronauts had completed 34 orbits in six missions, while the cosmonauts had completed 259 orbits in six missions.

Below: **The first woman in space, Soviet Vostock 6 cosmonaut Valentina Tereshkova, completed 70 hours and 50 minutes in space after liftoff on 16 June 1963.** *At right:* **The TIROS 8 weather satellite, part of NASA's Television Infrared Observation Satellite global 24-hour weather-watch net. TIROS 8 was the first satellite to test automatic picture transmission.**

At left: Identical Syncoms 1 and 2 were the first communications satellites ever to be placed in geosynchronous orbit. *Below:* A Hughes aircraft technician inspects a Syncom satellite's attitude motor. Explorer 18 *(above)* measured radiation, magnetic fields and solar wind, and discovered a region of high-energy radiation beyond the Van Allen belt. See the illustration *at right.*

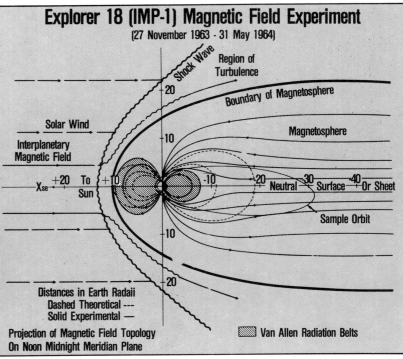

Explorer 18 (IMP-1) Magnetic Field Experiment
(27 November 1963 - 31 May 1964)

Distances in Earth Radaii
Dashed Theoretical ---
Solid Experimental —

Projection of Magnetic Field Topology
On Noon Midnight Meridian Plane

▨ Van Allen Radiation Belts

Space Flight Highlights of 1963

	Launch date	Launch vehicle	Launch weight
Syncom 1 (USA)	13 Feb	Thor-Delta	86 lb
Mercury 9 *(Faith 7)* (USA)	15 May	Atlas D	3000 lb
Vostok 5 *(Hawk)* (USSR)	14 Jun	A-1	10,384 lb
Vostok 6 *(Sea Gull)* (USSR)	16 Jun	A-1	10,369 lb
Explorer 18 (USA)	26 Nov	Delta	138 lb
TIROS 8 (USA)	21 Dec	Delta	297 lb

American Astronaut

6.	L Gordon Cooper	(Mercury 9)	15 May

Soviet Cosmonauts

5.	Valery Bykovsky	(Vostok 5)	14 Jun
6.	Valentina Tereshkova	(Vostok 6)	16 Jun

1964

In a year that saw a record number of NASA spacecraft launches, the Americans for the first time in four years placed no astronauts into space. Highlights of the unmanned NASA effort included the 28 July launch of Ranger 7, which returned 4316 high-resolution close-up television pictures of the moon's surface before it crashed into the Sea of Clouds three days later. Syncom 3, launched into geosynchronous orbit over the Pacific on 19 August, transmitted live television pictures of the Tokyo Olympics, while Mariner 4 was launched on 28 November on a successful seven-month journey to the planet Mars (*see 1965*).

Manned space flight in 1964 consisted only of the Soviet Voskhod 1 mission. Noting that the Americans were planning a series of two-man spacecraft missions in 1965, the Soviets were determined to be the first with a multiplace spacecraft. They had been the first to put unmanned, manned and womaned spacecraft into orbit and intended to be the first with a 'multimanned' mission. Consequently, Voskhod 1, which was launched on 12 October, was simply a greatly modified Vostok spacecraft which was reconfigured to carry three men on what would be a crowded 16 orbit, 24 hour flight, in which space suits were not used in the interest of reducing the crowding.

Below, left to right: **Commander Vladimir Komarov, physician Boris Yegorov and scientist Konstantin Feoktistov are here shown before their historic three-man flight aboard Voskhod 1. NASA's Mariner 4 *(at right)* left Earth on 28 November 1964 to eventually photograph part of the surface of Mars.**

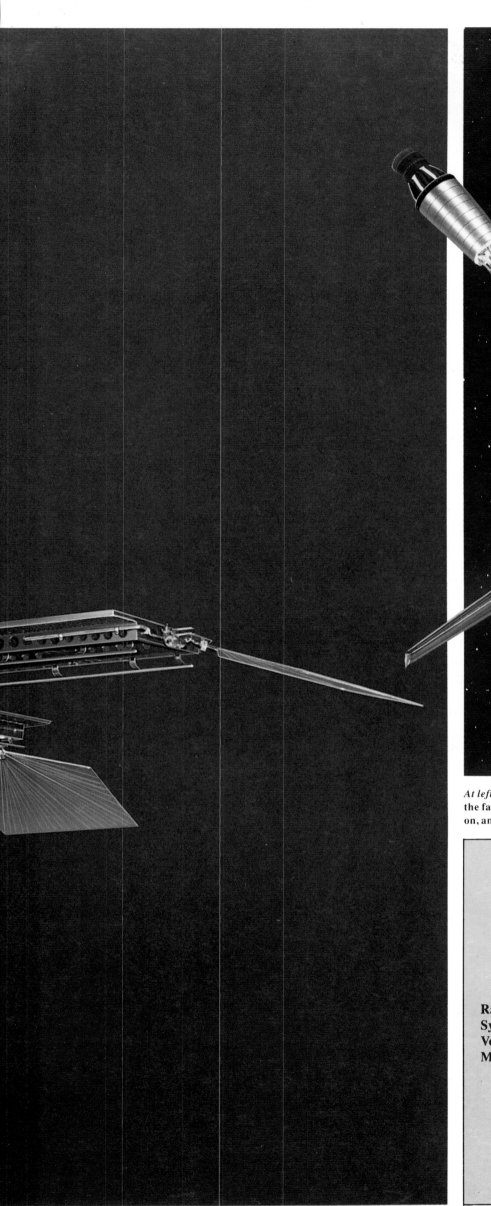

At left and above: The basic design of the Ranger spacecraft did not change greatly, yet the failures and successes of the various Rangers led up to the Ranger 7—which landed on, and photographed, the Moon.

Space Flight Highlights of 1964

	Launch date	Launch vehicle	Launch weight
Ranger 7 (USA)	28 Jul	Atlas/Agena	806 lb
Syncom 3 (USA)	19 Aug	Thor-Delta	86 lb
Voskhod 1 *(Ruby)* (USSR)	12 Oct	A-2	11,704 lb
Mariner 4 (USA)	28 Nov	Atlas/Agena D	575 lb

Soviet Cosmonauts
7. Vladimir Komarov   (Voskhod 1)   12 Oct
8. Konstantin Feoktistov
9. Boris Yegorov

1965

This was the year that can be remembered as the one in which the United States took the lead in the space race, a lead it was to hold for at least the next ten years until it temporarily abandoned manned space flight for six years in 1975.

The Soviet Union fielded the opening act of 1965 on 18 March with the second and last of the Voskhod series. This time the capsule carried only two cosmonauts. The flight's high point came when Alexei Leonov climbed out of Voskhod and became the first human being to walk in space. The low point of Voskhod 2 came when the automatic reentry system failed and the spacecraft landed in rugged Ural mountain country. The crew, which also included Pavel Belyayev, was forced to spend their first night back on earth in a windswept snowbank waiting to be rescued, but other than that, they were no worse for wear.

From then on, however, it was America's year in space. The first flight of the two-man Gemini spacecraft series came on 23 March with America's second man in space, Gus Grissom, teamed with John Young for a three orbit flight. Young was destined to be the only man to fly six space flights in three different types of spacecraft—Gemini, Apollo and Space Shuttle—completing two flights in each type of vehicle over an 18 year period.

Interspersed within the fast-paced series of Gemini flights was the 6 April launch of Intelsat 1, aka *Early Bird,* the first private industry developed (Comsat Corporation) geosynchronous satellite; four Explorer launches; and the successful flights of Ranger 8 and Ranger 9, which both reached the moon and returned a total of 12,951 photographs of the lunar surface before impact.

Mariner 4, launched in 1964, came within 6118 miles of the planet Mars on 14 July, marking a major milestone in early planetary exploration and returning the first ever close-up photos of the red planet.

The summer of 1965 saw the launch of the second two of the manned Gemini spacecraft. James McDivitt and Ed White were launched on 3 June for a 62 orbit flight in which White became the first American to walk in space. Mercury veteran Gordon Cooper was joined by Charles Conrad for the 120 orbit flight of Gemini 5, which was launched on 21 August.

The two major achievements of the Gemini program both came in December during the simultaneous flights of Gemini 6 and Gemini 7. Soviet spacecraft had flown simultaneous space flights during the Vostok program, but the two spacecraft were widely spaced and did not interact with one another. Gemini 6, however, launched on 15 December, was piloted by Wally Schirra and Tom Stafford to a rendezvous with Frank Borman and James Lovell in Gemini 7, which had been launched on 4 December. The second major accomplishment by the Gemini program in December 1965 was the fact that Borman and Lovell remained in orbit for an unprecedented *two weeks,* completing 206 orbits and establishing a space duration record that would stand until the deployment of the American Skylab space station in 1973. Their 330.6 hours in space was a longer time spent aboard spacecraft than any of the Apollo lunar flights by more than a day!

Having established a national space agency (*Centre National d'Etudes Spatiales,* CNES), France, in 1965, became the third nation to launch a satellite. Launched from Hammaguir, a military base in the Algerian desert, on 26 November, the first French spacecraft was the A-1 *Asterix.* France would launch her second satellite, D-1A *Diapason,* in February 1966, and would go on to become the world's third most prolific space power by a large margin.

Below, left and right: **Mission commander James A McDivitt and pilot Edward H White II train in a Gemini capsule simulator previous to their flight aboard Gemini 4.** *At right:* **A Gemini capsule is hoisted to its Titan booster.**

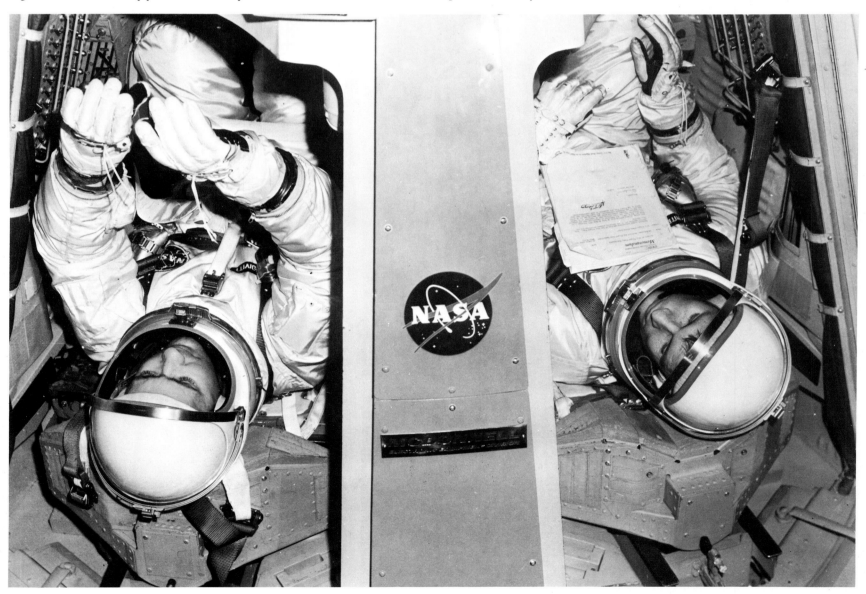

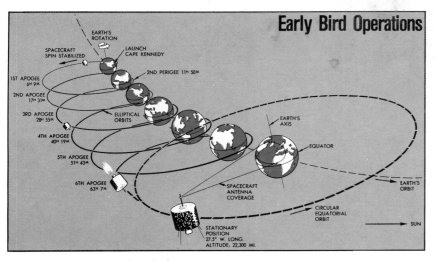

Early Bird Operations

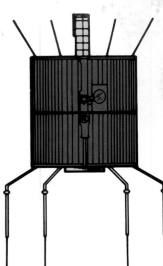

At right and below: The first private-industry developed geosynchronous satellite, Comsat's Intelsat 1, aka 'Early Bird,' and *(left)* how it achieved its Earth-stationary orbit. *Opposite:* Intelsat 1 receives an inspection in the lab; satellite 'white rooms' are kept squeaky clean.'

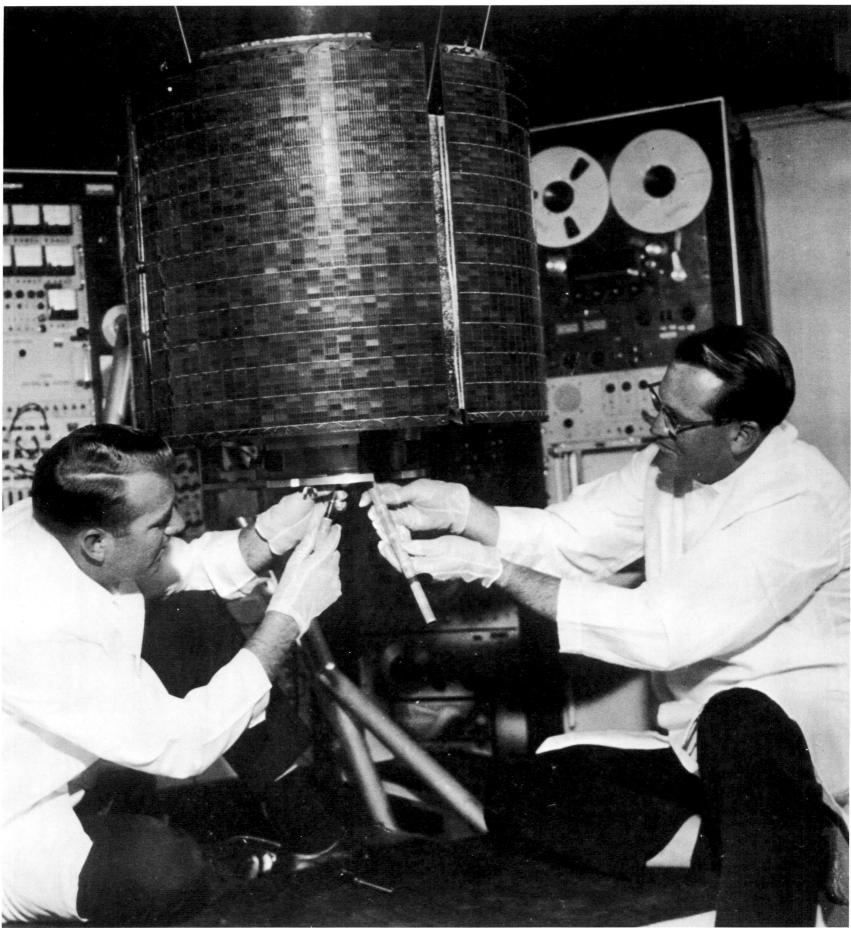

At left: The somewhat cozy interior of a Gemini capsule, with at least some systems 'on.' *Above:* Gemini is boosted into the sky by a Titan.

Space Flight Highlights of 1965

	Launch date	Launch vehicle	Launch weight
Voskhod 2 *(Diamond)* (USSR)	18 Mar	A-2	12,382 lb
Gemini 3 *(The Unsinkable Molly Brown)* (USA)	23 Mar	Titan 2	8200 lb
Intelsat 1/*Early Bird* (USA)	6 Apr	TAD	85 lb
Gemini 4 (USA)	3 Jun	Titan 2	8200 lb
Gemini 5 (USA)	21 Aug	Titan 2	8200 lb
A-1 *Asterix* (France)	26 Nov	Diamant	92 lb
Gemini 7 (USA)	4 Dec	Titan 2	8200 lb
Gemini 6 (USA)	15 Dec	Titan 2	8200 lb

American Astronauts

2.	Virgil Grissom (2nd flight)	(Gemini 3)	23 Mar
7.	John Young		
8.	James McDivitt	(Gemini 4)	3 Jun
9.	Edward White		
6.	L Gordon Cooper (2nd flight)	(Gemini 5)	21 Aug
10.	Charles Conrad		
11.	Frank Borman	(Gemini 7)	4 Dec
12.	James Lovell		
5.	Walter Schirra (2nd flight)	(Gemini 6)	15 Dec
13.	Thomas Stafford		

Soviet Cosmonauts

10.	Pavel Belyayev	(Voskhod 2)	18 Mar
11.	Alexei Leonov		

1966

It was the first year since Yuri Gagarin's Vostok 1 that *no* Soviet cosmonaut went into space. In the meanwhile, ten American astronauts flew five Gemini missions and both countries made major strides in Lunar exploration.

On 31 January, the Soviet Luna 9 was launched, and on 3 February it became the first spacecraft to reach the moon intact, landing in the Ocean of Storms and sending back a series of television pictures of the surface. On 31 March Luna 10 followed her sister to the moon where she became the first spacecraft to be placed into orbit around that body.

American lunar exploration activities directly paralleled those of the Soviets, though they lagged behind by several months. Surveyor 1 was launched on 30 May and successfully soft landed, like Luna 9, in the Ocean of Storms. The first of the appropriately named Lunar Orbiter spacecraft left the Earth on 10 August. Its orbital mission was to photograph potential manned landing sites on the moon's surface. An attempt to land a second surveyor on the moon in September failed, but another Lunar Orbiter successfully took up its station in November.

The objectives of the American Gemini manned spacecraft program for 1966 also included preparations for the projected manned lunar landing that was scheduled to come with the impending Apollo project. An important part of an Apollo flight was to involve the manned Apollo command Module docking with the Apollo Lunar Module that would actually carry the landing crew to the moon's surface. In order to prepare for this maneuver, Gemini crews practiced docking with an Agena rocket upper stage that was launched from an adjoining launch pad at roughly the same time as the Gemini craft.

In the first such mission in March, Neil Armstrong and David Scott aboard Gemini 8 successfully docked with their target but had to break off earlier than expected due to a thruster malfunction. In June, Gene Cernan and Tom Stafford rendezvoused successfully with their target but couldn't dock with it because its shroud failed to drop off the Agena's docking adapter.

The following month, Gemini 10 with John Young and Michael Collins rendezvoused with both their own target and that of Gemini 8 as well. Having successfully docked with the former, they actually used its propellant for maneuvering the pair of docked spacecraft. Both the Gemini 11 and Gemini 12 crews in September and November successfully docked with their targets within their first three orbits of the earth. Thus, the Gemini program concluded after ten successful flights which during 1966 had seen six successful rendezvous on six attempts and four successful docking maneuvers in five tries. As the year ended, the first of the new three-man Apollo lunar spacecraft had been delivered to the Kennedy Space Center and the Americans looked hopefully toward the moon.

Left: Surveyor 1 rises on an Atlas/ Centaur toward its successful soft Moon landing. *Above right:* The Agena target vehicle, the object of both success and failure for NASA's early space docking projects. *Below right:* Astronauts Eugene Cernan and Thomas Stafford after the first, aborted, attempt to launch Gemini 9.

Space Flight Highlights of 1966

	Launch date	Launch vehicle	Launch weight
Luna 9 (USSR)	31 Jan	A2e	3384 lb
Gemini 8 (USA)	16 Mar	Titan 2	8200 lb
Luna 10 (USSR)	31 Mar	A2e	3520 lb
Surveyor 1 (USA)	30 May	Atlas/Centaur	2194 lb
Gemini 9 (USA)	3 Jun	Titan 2	8200 lb
Gemini 10 (USA)	18 Jul	Titan 2	8200 lb
Lunar Orbiter 1 (USA)	10 Aug	Atlas/Agena	853 lb
Gemini 11 (USA)	12 Sep	Titan 2	8200 lb
Gemini 12 (USA)	11 Nov	Titan 2	18,297 lb

American Astronauts

14.	Neil Armstrong	(Gemini 8)	16 Mar
15.	David Scott		
13.	Thomas Stafford (2nd flight)	(Gemini 9)	3 Jun
16.	Eugene Cernan		
7.	John Young (2nd flight)	(Gemini 10)	18 Jul
17.	Michael Collins		
10.	Charles Conrad (2nd flight)	(Gemini 11)	12 Sep
18.	Richard Gordon		
12.	James Lovell (2nd flight)	(Gemini 12)	11 Nov
19.	Edwin Aldrin		

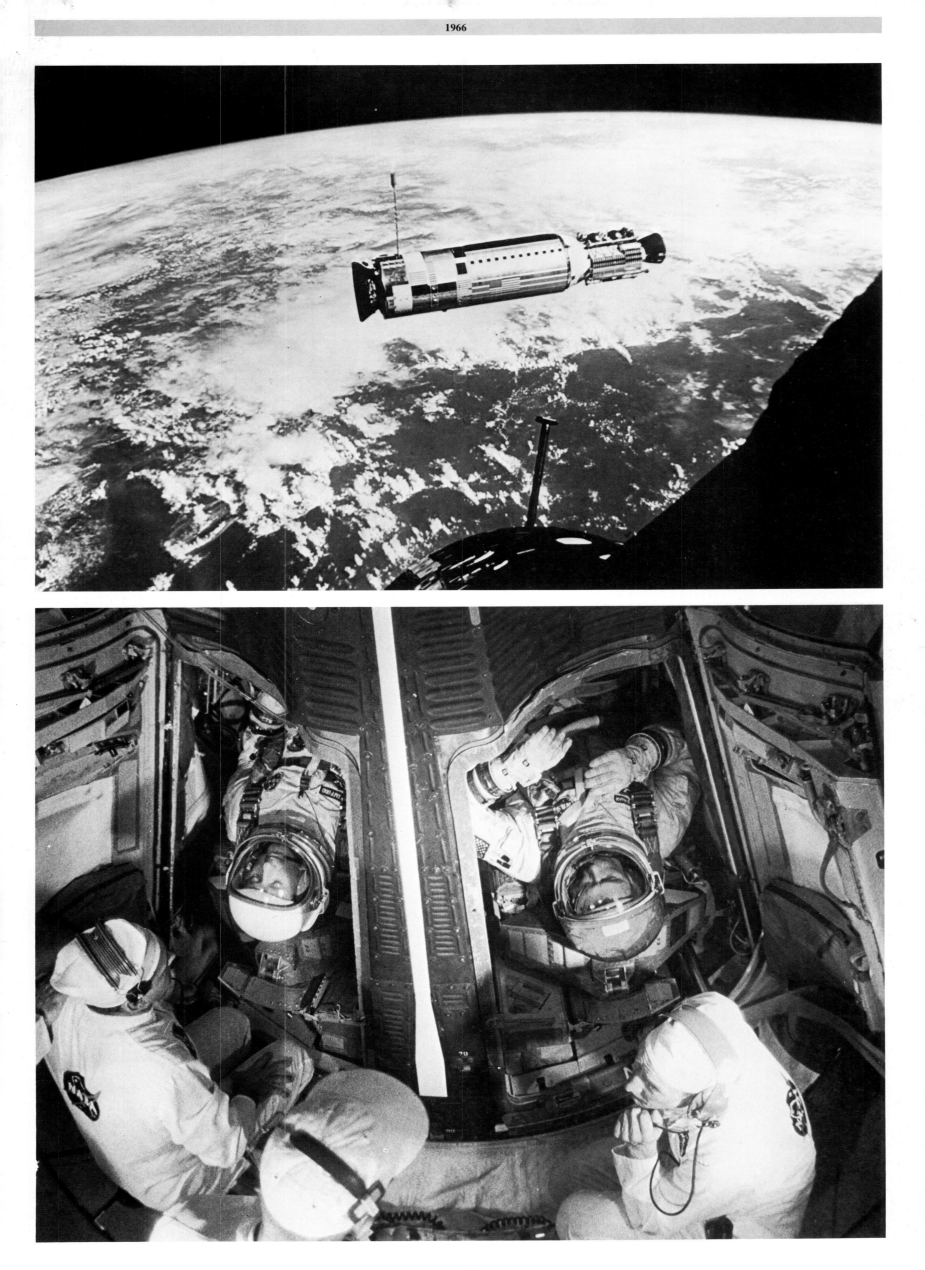

Part Three

The Race To The Moon

1967–1972

1967

A year of hope gave way to a year of disaster for the manned space programs of both the United States and the Soviet Union. On 27 January astronauts Gus Grissom, Ed White and Roger Chaffee were aboard the Apollo 1 spacecraft practicing for their 21 February launch when an electrical fire suddenly started and quickly engulfed the 100% oxygen atmosphere within the spacecraft. All three men perished before the capsule could be opened and the fire extinguished. The American manned space program was put on hold, and the three Apollo flights scheduled for 1967 were postponed indefinitely as the spacecraft went back for a complete redesign.

Like the Americans, the Soviets also intended to unveil a new three-man spacecraft in 1967, and like the Americans, the Soviets were also visited by the hand of violent death. On 23 April Vladimir Komarov was launched on what proved to be a routine flight aboard Soyuz 1—routine, that is, until reentry. As he was descending for a landing on 24 April the parachute lines above Komarov's craft became tangled, the parachute failed, and he crashed to his death.

Neither country conducted another manned launch during 1967, but unmanned launches continued unabated. The United States orbited three more Lunar Orbiters for three attempts and successfully soft landed three more Surveyor spacecraft on the moon in four tries. Among other spacecraft, the American Explorer series had progressed through Explorer 36 and the Soviet Union launched its Cosmos 188.

France, the world's third space power, meanwhile launched two spacecraft during February 1967: D-1C (*Diademe* 1) and D-1 (*Diademe* 2). On 29 November, Australia became the world's fourth space power by launching her tiny Wresat (Weapons Research Establishment Satellite) from the Woomera launch facility.

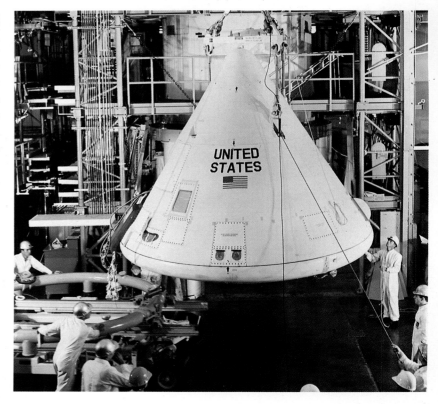

Above: The three-man Apollo spacecraft. *Below left:* Gus Grissom, Ed White and Roger Chaffee died in the tragic Apollo 1 blaze. *Below:* Yet another Surveyor heads for the Moon in the Surveyor-intensive year of 1967. *At right:* The Surveyor spacecraft paved the way for manned Moon missions.

Space Flight Highlights of 1967

	Launch date	Launch vehicle	Launch weight
Soyuz 1 (USSR)	23 Apr	A-2	14,344 lb
Lunar Orbiter 4 (USA)	4 May	Atlas/Agena	853 lb
Venara 4 (USSR)	12 Jun	A-2e	2212 lb
Explorer 35 (USA)	19 Jul	Delta	230 lb
Surveyor 6 (USA)	7 Nov	Atlas/Centaur	2290 lb
Wresat (Australia)	29 Nov	Sparta	99 lb

Soviet Cosmonaut

7. Vladimir Komarov (Soyuz 1) 23 Apr
(2nd flight)

Surveyor 6 (USA)

SOLAR PANEL

OMNIDIRECTIONAL ANTENNA A

HIGH-GAIN ANTENNA

THERMALLY CONTROLLED COMPARTMENT B

TV CAMERA

THERMALLY CONTROLLED COMPARTMENT A

STAR CANOPUS SENSOR

RADAR ALTITUDE-DOPPLER VELOCITY SENSOR

OMNIDIRECTIONAL ANTENNA B

FOOTPAD 3

FOOTPAD 2

VERNIER ENGINE 3

CRUSHABLE BLOCK

VERNIER PROPELLANT PRESSURIZING GAS TANK (HELIUM)

SOIL SAMPLER

AUXILIARY BATTERY

RETRO ROCKET NOZZLE

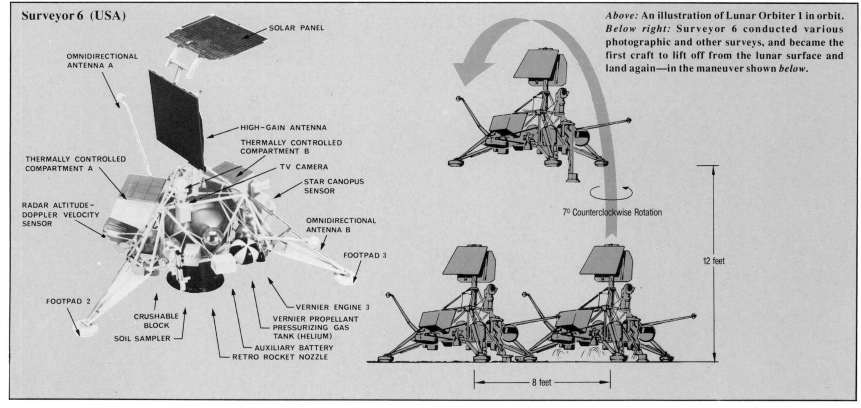

7⁰ Counterclockwise Rotation

12 feet

8 feet

Above: An illustration of Lunar Orbiter 1 in orbit. *Below right:* Surveyor 6 conducted various photographic and other surveys, and became the first craft to lift off from the lunar surface and land again—in the maneuver shown *below.*

Venera 4 (USSR)

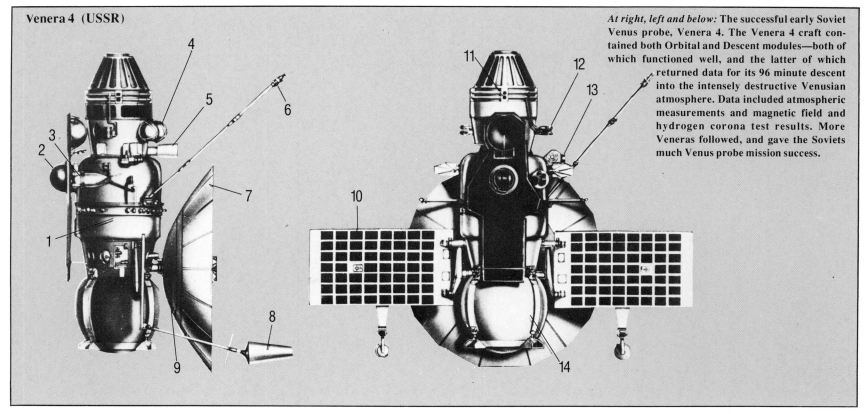

At right, left and below: The successful early Soviet Venus probe, Venera 4. The Venera 4 craft contained both Orbital and Descent modules—both of which functioned well, and the latter of which returned data for its 96 minute descent into the intensely destructive Venusian atmosphere. Data included atmospheric measurements and magnetic field and hydrogen corona test results. More Veneras followed, and gave the Soviets much Venus probe mission success.

1968

The manned space programs in the Soviet Union and the United States each heated up during 1968 as the two countries raced toward the goal of a man on the moon before the end of the decade.

The first manned Apollo launch, Apollo 7, came on 11 October, 21 months after the Apollo 1 disaster of 1967. The crew, Wally Schirra, Donn Eisele and Walter Cunningham, spent 11 days in earth orbit during which time they successfully rendezvoused with the Saturn launch vehicle's upper stage.

The Soviet Union launched the unmanned Soyuz 2 on 25 October, which was followed the next day by Soyuz 3 with a single crewman, Giorgi Beregovoi, aboard. Beregovoi used Soyuz 2 successfully as a rendezvous target much in the same manner as the Gemini flights of 1966, but he made no attempt at docking. Meanwhile, the Soviets were conducting a series of unmanned flights to the moon with a series of Soyuz-type spacecraft designated Zond. Zond 5, launched on 14 September, reached the moon, looped around it and returned to earth where it was successfully recovered. The maneuver was successfully repeated in November with Zond 6 and the Soviet Union appeared as though it might be poised to do it again with a manned Soyuz spacecraft.

On December 21, however, the United States launched the Apollo 8 spacecraft atop the gargantuan Saturn 5, the largest rocket ever launched into space. The destination for the crew of three—Frank Borman, James Lovell and William Anders—was a rendezvous with the moon! They reached the moon, the first men to do so, on Christmas eve and put Apollo 8 into orbit around it for ten revolutions before coming home. As 1968 ended, no human had yet set foot on the lunar surface, but three sets of human eyes had seen the moon's far side for the first time and had watched the earth rise over the lunar horizon!

Below opposite: The Command Service Module for the Apollo 7 orbital flight is here being positioned for mating with its Saturn booster adaptor. *Below:* This, man's first view of 'Earthrise,' greeted the Apollo 8 astronauts as they came from behind the Moon following orbital insertion burn. The Sun-illuminated foreground features are on the Moon's eastern limb.

At left: The fifth NASA TIROS Operational Satellite, TOS-E, is here being prepared for launch in August, 1968. Once in orbit, this meteorological satellite, with its special weather surveillance gear, became ESSA-7 (Environmental Science Services Administration-7). *Below:* A closer look at a TOS/ESSA satellite is provided by this testbed view. TOS/ESSA satellites engendered the later ITOS (Improved TIROS Operational System) weather satellites. *Above, right to left:* Astronauts Walter M 'Wally' Schirra, Donn F Eisele and Walter Cunningham were photographed during Apollo 7 water egress training in the Gulf of Mexico. Apollo 7 completed 11 Earth orbits on 11 October 1968 for a mission duration of some 260 hours and eight minutes.

Space Flight Highlights of 1968

	Launch date	Launch vehicle	Launch weight
ESSA 7 (TOS-E) (USA)	16 Aug	Delta	320 lb
Apollo 7 (USA)	11 Oct	Saturn 4	62,389 lb (CSM)
Soyuz 3 (USSR)	26 Oct	A-2	14,465 lb
Apollo 8 (USA)	21 Dec	Saturn 5	63,650 lb (CSM)

American Astronauts

5.	Walter Schirra (2nd flight)	(Apollo 7)	11 Oct
20.	Donn Eisele		
21.	Walter Cunningham		
11.	Frank Borman (2nd flight)	(Apollo 8)	21 Dec
12.	James Lovell (3rd flight)		
22.	William Anders		

Soviet Cosmonaut

12.	Giorgi Beregovoi	(Soyuz 3)	26 Oct

1969

Whatever else can be said about 1969—within or without the annals of space flight—it will always be remembered as the year that men finally set their feet on lunar soil.

The year of the moon began with the Soviet launch of Soyuz 4 and Soyuz 5 on 14 and 15 January. The mission, designed in part to upstage Apollo 8, involved not only a rendezvous and the first docking between two manned spacecraft, but crew exchange as well. This is how it worked: Vladimir Shatalov went up alone in Soyuz 4, while Boris Volynov, Aleksei Yeliseyev and Yevgey Khrunov followed in Soyuz 5 the next day. After rendezvous and dock, Khrunov and Yeliseyev spacewalked to Soyuz 4 and returned to earth with Shatalov, while Volynov returned alone.

On March 3, the United States launched Apollo 9. The purpose of the flight was to test the Apollo Lunar Module (LM) in space for the first time. During the ten-day earth orbit mission James McDivitt and Russell Schweickart left the Apollo 9 Command and Service Modules (CSM), climbed aboard the LM, detached it, flew away and returned, then docked again with the CSM, which was flown by David Scott.

On 21 May Apollo 10—with Tom Stafford, John Young and Gene Cernan—became the second manned spacecraft to achieve lunar orbit. Having done so, they repeated the test that had been conducted by the Apollo 9 crew in earth orbit. Stafford and Cernan climbed aboard the LM and actually flew it to within 50,000 feet of the lunar surface before returning to a rendezvous with the CSM.

Continued on page 54

The Apollo Earth orbital test missions going up and coming down: Apollo 9 *(above)* **blasts off with a massive Saturn booster; and** *(below)* **the Apollo 10 Command Module parachutes down from a successful test of lunar mission apparatus.** *At right:* **Apollo 9's Lunar Landing Test Module in Earth orbit.**

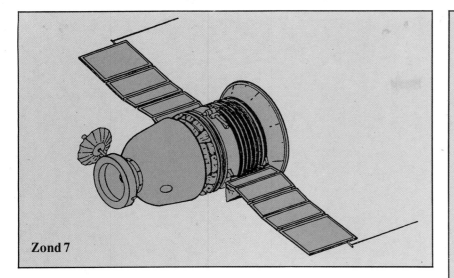

Zond 7

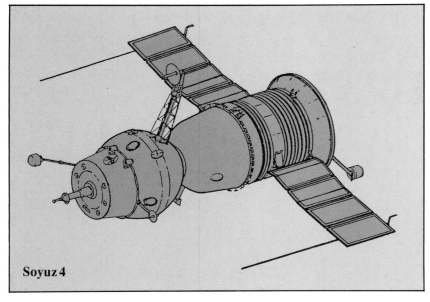

Soyuz 4

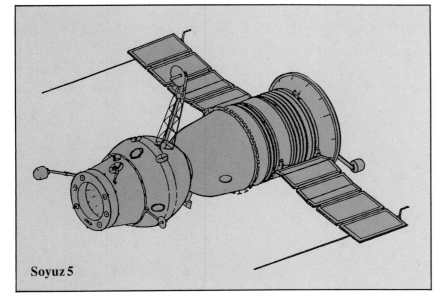

Soyuz 5

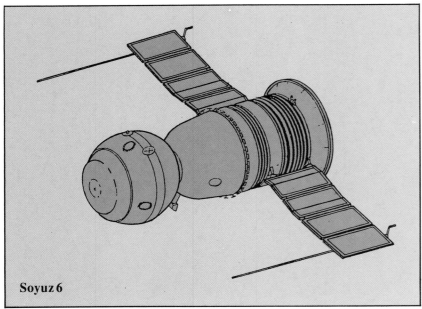

Soyuz 6

Space Flight Highlights of 1969

	Launch date	Launch vehicle	Launch weight
Soyuz 4 (USSR)	14 Jan	A-2	14,575 lb
Soyuz 5 (USSR)	15 Jan	A-2	14,487 lb
Apollo 9 (USA)	3 Mar	Saturn 5	48,564 lb (CSM) (*Gumdrop*)
Apollo 10 (USA)	18 May	Saturn 5	63,648 lb (CSM) (*Charlie Brown*)
			30,849 lb (LM) (*Snoopy*)
Apollo 11 (USA)	16 Jul	Saturn 5	63,493 lb (CSM) (*Columbia*)
			33,205 lb (LM) (*Eagle*)
Zond 7 (USSR)	7 Aug	D-1e	11,825 lb
Soyuz 6 (USSR)	11 Oct	A-2	14,469 lb
Soyuz 7 (USSR)	12 Oct	A-2	14,621 lb
Soyuz 8 (USSR)	13 Oct	A-2	14,621 lb
Apollo 12 (USA)	14 Nov	Saturn 5	63,470 lb (CSM) (*Yankee Clipper*)
			33,325 lb (LM) (*Intrepid*)

American Astronauts

8.	James McDivitt (2nd flight)	(Apollo 9)	3 Mar
15.	David Scott (2nd flight)		
23.	Russell Schweickart		
13.	Thomas Stafford (3rd flight)	(Apollo 10)	18 May
16.	Eugene Cernan (2nd flight)		
7.	John Young (3rd flight)		
14.	Neil Armstrong (2nd flight)	(Apollo 11)	16 Jul
19.	Edward Aldrin (2nd flight)		
17.	Michael Collins (2nd flight)		
10.	Charles Conrad (3rd flight)	(Apollo 12)	14 Nov
18.	Richard Gordon (2nd flight)		
24.	Alan Bean		

Soviet Cosmonauts

13.	Vladimir Shatalov	(Soyuz 4)	14 Jan
14.	Boris Volynov	(Soyuz 5)	15 Jan
15.	Yevgeny Khrunov		
16.	Alexei Yeliseyev		
17.	Georgi Shonin	(Soyuz 6)	11 Oct
18.	Valeri Kubasov		
19.	Anatoli Filipchenko	(Soyuz 7)	12 Oct
20.	Vladislav Volkov		
21.	Viktor Gorbatko		
13.	Vladimir Shatalov (2nd flight)	(Soyuz 8)	13 Oct
16.	Alexei Yeliseyev (2nd flight)		

Opposite: Carrying mankind to his first walk on the Moon, Apollo 11 lifts off, its massive Saturn 5 thrusters blazing a path into history. *At left, from the top down:* Soviet spacecraft: Zond 7 took high quality color photos of the Earth and the Moon, on a seven day journey in August, 1969; Soyuz 4 and 5 conducted the first docking of manned spacecraft in orbit; and Soyuz 6 participated in Earth-orbital, manned formational flying with Soyuz 7 and 8.

Continued from page 50

Centuries of dreams came true with the flight of Apollo 11 which was launched on 16 July 1969 with Neil Armstrong, Michael Collins and Edwin Aldrin aboard. On 20 July, Armstrong and Collins took the Apollo 11 LM to a historic landing in the moon's Sea of Tranquillity. They spent a total of 21.6 hours on the lunar surface before returning to the CSM and thence to a successful landing back on the earth on 24 July.

The Soviet Union also sent a spacecraft to the moon in July 1969, but the unmanned Luna 15 was hardly noticed when it went out of control and crashed, appropriately, into the Sea of Crises as Armstrong and Collins were preparing to lift off the lunar surface.

Between 11 and 13 October the Soviet Union launched Soyuz 6, Soyuz 7 and Soyuz 8, with a total of seven cosmonauts. Each spacecraft remained in orbit for five days with three days of overlap during which they conducted some elaborate formation flying. The three ships, however, remained in earth orbit as the Soviet Union had abandoned its manned lunar landing effort with the comment that it had never intended to go there in the first place.

For its part, the United States ended the decade with the almost anticlimatic *second* lunar landing conducted by Charles Conrad and Alan Bean in the Apollo 12 LM in November.

'That's one small step for man, one giant leap for mankind.' Apollo 11 rises past the launch tower at pad 39A on 16 July 1969 *(at top of page)*, **and four days later, Neil Armstrong made man's first footprint on the Moon** *(below)*. *At right:* **The Apollo 12 Lunar Module descends into the Moon's Ocean of Storms.**

1970

If 1969 had been a year of unparalleled triumph for the United States, 1970 saw a near disaster that harkened back to the gloom that overshadowed 1967. Apollo 13 was launched on 11 April with James Lovell, John Swigert and Fred Haise bound for a third lunar landing. All of the optimism surrounding the Apollo program's 1969 successes vanished in a flash as an explosion occurred in one of Apollo 13's oxygen tanks on 13 April. Though no one was hurt in the explosion, a great deal of oxygen was lost and it was too late to turn back without going to the moon and using its gravity to slingshot the spacecraft back to earth. There was a very real possibility that the crew would be lost in space, but thanks in part to their own ingenuity of improvisation, they survived and returned to a successful landing back on the earth on 17 April. The lunar landing was, of course, aborted and Apollo 14 was postponed to 1971.

1970 was the year that Japan and China became the fifth and sixth nations to successfully orbit a satellite. On 11 February the experimental Osumi spacecraft was launched from the Japanese Kagoshima launch center on the island of Kyushu. Osumi's transmitters broadcast for a total of seven orbits, but the spacecraft—really an instrument-packed Lambda fourth stage—remained in orbit. The first Chinese satellite, designated simply 'China 1,' was launched on 24 April, whereupon it orbited the earth for two months broadcasting the anthem 'The East is Red'.

Having officially given up on a manned lunar landing, the Soviet Union proceeded in June with Soyuz 9, a manned earth orbital mission in which Andrian Nikolayev and Vitali Sevastyanov set a new space duration record of 425 hours.

Having conceded one phase of the race to the moon, the Soviet Union contrived and successfully executed a pair of spectacular and quite unique lunar expeditions of their own. On 12 September, they launched the robot probe Luna 16, which conducted a landing on the moon, collected 22 pounds of soil samples, blasted off, and returned the samples to earth on 21 September. It was the first ever remote control sample return from another celestial body. The Luna 16 mission was followed by the equally intriguing flight of Luna 17. Launched on 10 November, this lunar lander carried Lunokhod 1, a robot lunar rover that looked somewhat like a cross between an insect and a toy.

Lunokhod traveled more than six miles across the moon's surface under remote control, sampling the lunar soil as it went.

Space Flight Highlights of 1970

	Launch date	Launch vehicle	Launch weight
Ohsumi (Japan)	11 Feb	Lambda	4S-553 lb
Apollo 13 (USA)	11 Apr	Saturn 5	63,426 lb (CSM) (*Odyssey*)
			32,124 lb (LM) (*Aquarius*)
'China 1' (China)	24 Apr	CZ-1	378.4 lb
Soyuz 9 (USSR)	1 Jun	A-2	14,300 lb
Luna 16 (USSR)	12 Sep	D-1e	3991 lb
Luna 17 (USSR)	10 Nov	D-1e	3991 lb

American Astronauts

12.	James Lovell	(Apollo 13)	11 Apr
	(4th flight)		
25.	John Swigert		
26.	Fred Haise		

Soviet Cosmonauts

3.	Andrian Nikolayev	(Soyuz 9)	1 Jun
	(2nd flight)		
22.	Vitali Sevastyanov		

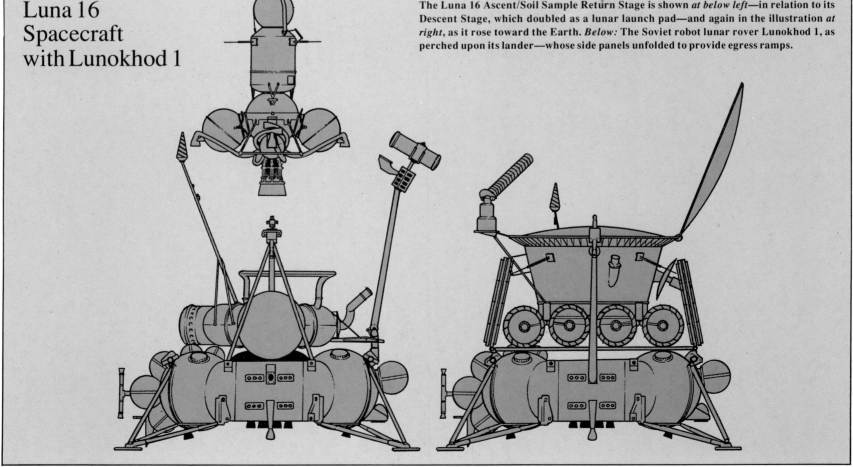

Luna 16 Spacecraft with Lunokhod 1

The Luna 16 Ascent/Soil Sample Return Stage is shown *at below left*—in relation to its Descent Stage, which doubled as a lunar launch pad—and again in the illustration *at right*, as it rose toward the Earth. *Below:* The Soviet robot lunar rover Lunokhod 1, as perched upon its lander—whose side panels unfolded to provide egress ramps.

1971

The United States made an early comeback in 1971 from the brush with disaster of Apollo 13 in April 1970. The long-awaited Apollo 14 Lunar landing mission was launched on 31 January. Commanded by Alan Shepard, who, ten years earlier, was America's first man in space, the Apollo 14 ran as smoothly as either of the successful lunar landings in 1969. Shepard and Edgar Mitchell spent more than nine hours outside the lander exploring the moon's surface. The launch of Apollo 15 followed on 26 July and once again the mission ran smoothly. The first of the three Apollo J-series missions, Apollo 15 was also the first to carry the Lunar Roving Vehicle, a small car that allowed astronauts David Scott and James Irwin to travel 17.5 miles over the lunar surface as they spent nearly twice the time outside the lunar lander as the Apollo 14 astronauts.

For the Soviet Union, 1971 became the year of Salyut as its first space station was placed into orbit in April. The crew of Soyuz 10, launched on 22 April, docked with Salyut but remained in the spacecraft without entering the space station. The Soyuz 11 cosmonauts, Georgi Dobrovolsky, Viktor Patsayev and Vladislav Volkov, were launched on 6 June and, having successfully docked with Salyut, they spent 22 days aboard the station. The triumph of this first ever operational use of a space station was tragically marred when the three men died during reentry into the earth's atmosphere after a valve opened accidentally and the spacecraft lost pressurization.

Soviet unmanned planetary exploration efforts during 1971 also went badly. The Mars 2 and Mars 3 spacecraft, both launched in May, reached the red planet on 27 November and 2 December respectively. Mars 2 crashed on the surface, and whereas Mars 3 succeeded in making a fairly soft landing, it had the misfortune of setting down in the midst of a dust storm which ruined its delicate instruments before it could record any useful data.

For the United Kingdom, 1971 was the year of the space race as they launched their first spacecraft, Prospero, from Woomera, Australia on 28 October.

Above: The biggest American launch vehicle, the Saturn 5, and its Apollo spacecraft cargo approach the launch site. *Left:* The Soviets' ill-fated Mars 3 probe. *At right:* The historic Apollo 15 Lunar Rover mission. Here the Command and Service Modules are as photographed—above the Moon—from the Lunar Lander.

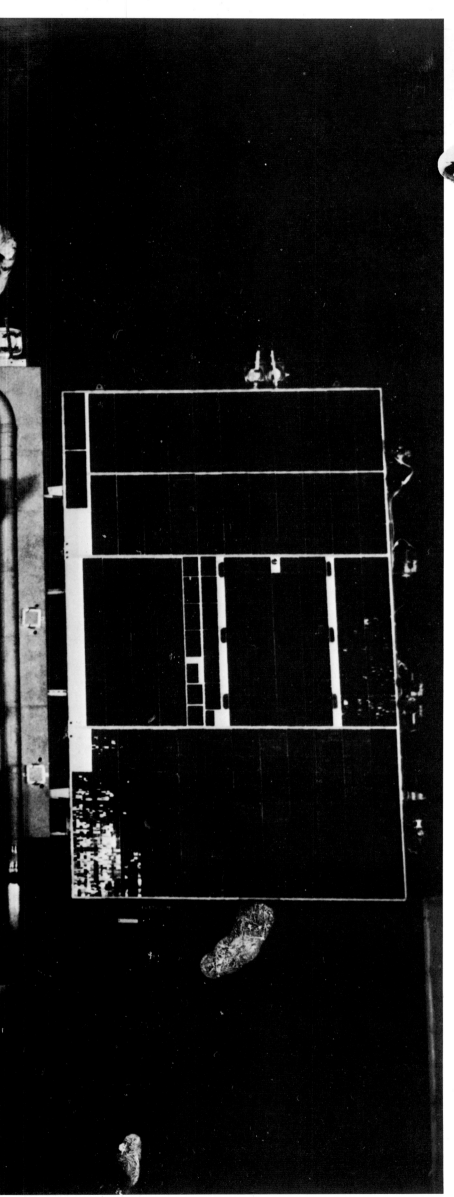

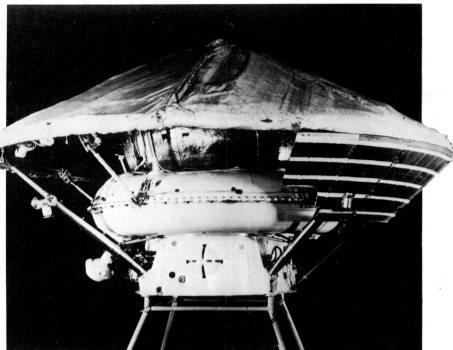

At left: The Soviet Mars 3 was identical to Mars 2. Here, the lander/orbiter combination is seen in a view complimentary to that on page 58. The Mars 3 Lander itself is in the isolated view *above*. The 'coolie hat' is a heat shield, which was not protection enough for Mars 3's delicate instruments, which were destroyed in a Martian dust storm. Though the Soviet Mars mission program continued into 1973–74, only one of seven Soviet Mars probes succeeded.

Space Flight Highlights of 1971

	Launch date	Launch vehicle	Launch weight
Apollo 14 (USA)	31 Jan	Saturn 5	64,438 lb (CSM) *(Kitty Hawk)*
			33,680 lb (LM) *(Antares)*
Salyut 1 (USSR)	19 Apr	D-1	40,260 lb
Soyuz 10 (USSR)	23 Apr	A-2	14,465 lb
Mars 3 (USSR)	28 May	D-1e	10,230 lb
Soyuz 11 (USSR)	6 Jun	A-2	14,938 lb
Apollo 15 (USA)	26 Jul	Saturn 5	66,818 lb (CSM) *(Endeavor)*
			36,230 lb (LM) *(Falcon)*
Prospero (UK)	28 Oct	Black Arrow	145 lb

American Astronauts
1.	Alan Shepard (2nd flight)	(Apollo 14)	31 Jan
27.	Stuart Roosa		
28.	Edgar Mitchell		
15.	David Scott (3rd flight)	(Apollo 15)	26 Jul
29.	James Irwin		
30.	Alfred Worden		

Soviet Cosmonauts
13.	Vladimir Shatalov (3rd flight)	(Soyuz 10)	23 Apr
16.	Alexei Yeliseyev (3rd flight)		
23.	Nikolai Rukavishnikov		
24.	Georgi Dobrovolsky	(Soyuz 11)	6 Jun
20.	Vladislav Volkov (2nd flight)		
25.	Viktor Patsayev		

These pages. Astronaut John Young, Commander of the Apollo 16 lunar mission, salutes the American flag as he leaps with ease in the Moon's comparatively weak gravity. The presence of the Lunar Rover, parked in front of the Lunar Module *Orion*, signifies that this was the second US Lunar Rover mission. Lunar Module pilot Charles M Duke took this picture.

1972

Both the United States and the Soviet Union took important steps in solar system exploration during 1972. For the Soviets it was Luna 20, launched on 14 February, which was their second successful robot sample return mission to the moon. Like Luna 16 two years before, the Luna 20 lander transmitted photographs and bored into the lunar surface with a rotary percussion drill.

On 2 March the United States launched Pioneer 10, the first spacecraft destined to explore the mysterious reaches of the outer solar system. Beyond the launch there was little to do but wait, for it would be 21 months before the probe reached its first stop—Jupiter.

It was in 1972 that the United States also made its two last Apollo lunar landings. Apollo 16, launched on 16 April, saw astronauts John Young and Charles Duke spend 71 hours on the lunar surface, including 20 hours outside the LM. Apollo 17, launched on 7 December, was the last of six successful lunar landings and the third to carry a Lunar Roving Vehicle. Having reached the moon, astronauts Gene Cernan and Harrison Schmitt spent 22 hours outside the spacecraft, covering 22 miles in the Lunar 'Rover'. The last human activity on the lunar surface during the Apollo program came at 5:23 pm EST on 13 December, with liftoff from the moon at 5:55 pm. The Apollo 17 crew returned to earth on 19 December, 310 hours and 51 minutes after they first blasted off, making the last Apollo lunar mission also the longest. They returned almost four years to the day after the launch of Apollo 8, the first spacecraft to travel to the vicinity of the moon.

Above: Apollo 16 Command and Service Modules in lunar orbit. *Below:* The last lunar mission, Apollo 17: Harrison Schmitt stows soil samples aboard the third Lunar Rover, near Shorty Crater (to photo right). *At right:* Man, the US flag and the Earth far above. Will we walk the Moon again?

At left: The Intelsat 4 F-5 satellite, launched by NASA for the international 'Intelsat' consortium. *Above:* The NASA Interplanetary Monitoring Platform satellite, IMP-H (aka Explorer 47). *At right:* Small Astronomy Satellite SAS-B (aka Explorer 48), launched by Italy for NASA on 16 November.

Space Flight Highlights of 1972

	Launch date	Launch vehicle	Launch weight
Luna 20 (USSR)	14 Feb	D-1e	3991 lb
Pioneer 10 (USA)	3 Mar	Atlas/ Centaur	570 lb
Apollo 16 (USA)	16 Apr	Saturn 5	66,918 lb (CSM) *(Casper)* 36,218 lb (LM) *(Orion)*
Intelsat 4 F-5 (USA)	13 Jun	Atlas/ Centaur	1587 lb
Explorer 47 (USA) (IMP-H)	22 Sept	Delta	860 lb
Explorer 48 (USA) (SAS-B)	16 Nov	Scout	410 lb
Apollo 17 (USA)	7 Dec	Saturn 5	66,844 lb (CSM) *(America)* 36,262 lb (LM) *(Challenger)*

American Astronauts
   7.   John Young   (Apollo 16)   16 Apr
         (4th flight)
  31.   Thomas Mattingly
  32.   Charles Duke
  16.   Eugene Cernan   (Apollo 17)   7 Dec
         (3rd flight)
  33.   Ronald Evans
  34.   Harrison Schmitt

NO SMOKIN

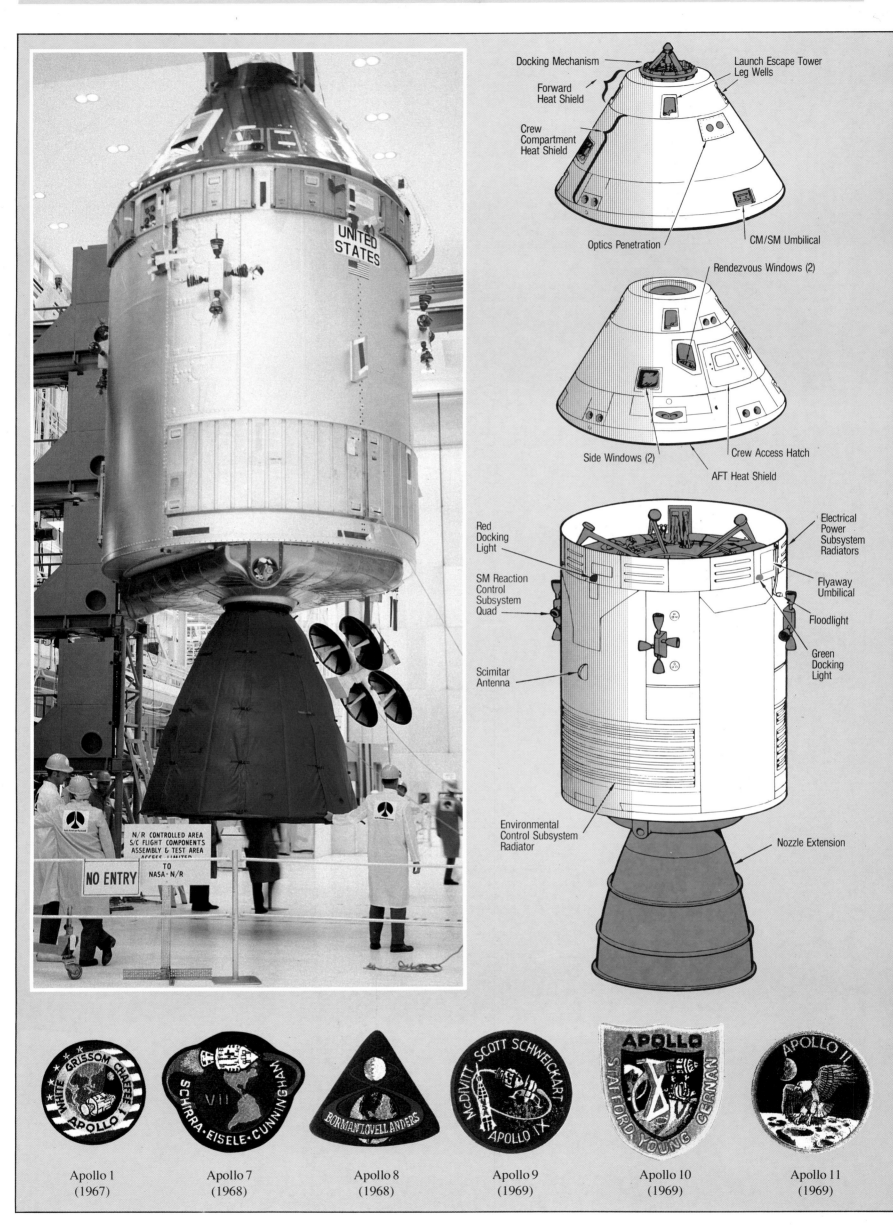

UNITED STATES

N/R CONTROLLED AREA
S/C FLIGHT COMPONENTS
ASSEMBLY & TEST AREA
ACCESS LIMITED
TO
NASA - N/R

NO ENTRY

Docking Mechanism

Launch Escape Tower
Leg Wells

Forward
Heat Shield

Crew
Compartment
Heat Shield

Optics Penetration

CM/SM Umbilical

Rendezvous Windows (2)

Side Windows (2)

Crew Access Hatch

AFT Heat Shield

Red
Docking
Light

Electrical
Power
Subsystem
Radiators

SM Reaction
Control
Subsystem
Quad

Flyaway
Umbilical

Floodlight

Scimitar
Antenna

Green
Docking
Light

Environmental
Control Subsystem
Radiator

Nozzle Extension

Apollo 1
(1967)

Apollo 7
(1968)

Apollo 8
(1968)

Apollo 9
(1969)

Apollo 10
(1969)

Apollo 11
(1969)

The Apollo Spacecraft

The hardware that supported the Apollo lunar missions between 1968 and 1972 was a system of some of the most sophisticated spacecraft ever designed. The centerpiece of the system was the Command/Service Module or CSM (*facing page*). The photo shows the Apollo 13 CSM (CSM-109) freshly delivered to Cape Kennedy from Rockwell International, with blue protective coating on the Command Module and special red padding on the Service Module's rocket engines. The diagram *on the facing page* shows (from top) the Command Module's outer shield, the Command Module (CM) itself and the Service Module (SM). The overall diameter of the CSM was 12 feet, 10 inches. It stood four stories high and weighed six tons.

The top of the Lunar Module (LM), seen *below*, was a hatch which connected to the tip of the CM during the flight to the Moon. Having arrived in lunar orbit, two of the three astronauts aboard the CSM would crawl through the hatch, detach the LM and land on the Moon as shown in the photo *above*. Having completed their lunar excursion, the two astronauts would lift off from the Moon's surface in the LM's Ascent Stage only. The Ascent Stage held crew quarters and all the LM's control and communications systems as well as the ascent rocket engine. The Descent Module (DM), with foot pads, contained the motor used in the descent to the Moon's surface and was always abandoned there by the astronauts. The Ascent Module was, in turn, abandoned in lunar orbit after returning the two astronauts to the CSM. The entire crew would return to Earth orbit in the CSM, abandon the SM in orbit, and would return to Earth in the CM only.

Lunar Rovers (*in the photo above*) were carried aboard the LMs of Apollo 15, 16 and 17 in 1971 and 1972. Rated at a maximum speed of 8.7 mph, one was actually driven (downhill) at a breathtaking 10.6 mph. They weighed 462 pounds and measured 72 x 137 inches. The three Rovers that were driven on the lunar surface logged a total of 56.1 miles.

Each Apollo spacecraft was launched into space by one of two launch vehicles. The Saturn 1B (*right*) was used for missions in Earth orbit, while the huge Saturn 5 (*far right*) was used for lunar missions. The Saturn 1B, whose rocket engines delivered two million pounds of thrust, stood 224 feet high. The Saturn 5, whose engines delivered *nine million* pounds of thrust, stood 363 feet high. The Saturn 5 still stands as the most powerful rocket ever used for space flight.

The badges shown *across the bottom of this spread* represent all Apollo lunar landing program missions with crews assigned. All went into space except Apollo 1, which was destroyed by fire. All others went to the Moon except Apollo 7 and 9, which were designated for Earth-orbit missions only. All missions from Apollo 11 forward landed on the Moon except Apollo 13, which aborted its mission en route because of an explosion in the SM. Apollo 17 was followed by four Apollo missions unrelated to lunar landing missions.

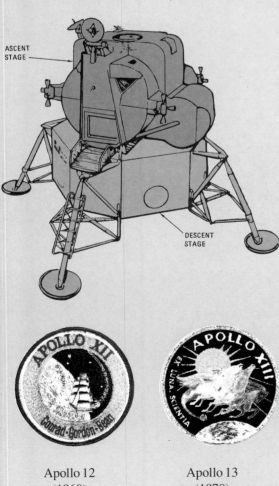

ASCENT STAGE

DESCENT STAGE

| Apollo 12 (1969) | Apollo 13 (1970) | Apollo 14 (1971) | Apollo 15 (1971) | Apollo 16 (1972) | Apollo 17 (1972) |

Part Four

The Age Of The Space Station

1973–1980

1973

Less than a month after the United States closed its book on lunar exploration, the Soviet Union returned to the moon with another robot explorer. Launched on 8 January, Luna 21 was the second spacecraft to carry a Lunokhod remotely controlled lunar rover. The spacecraft landed just 112 miles from where Apollo 17 had been the month before. Lunokhod 2 transmitted data for more than three months and traveled 23 miles, twice the distance covered by Lunokhod 1 in 1970.

The Soviet Union also launched two pairs of Mars probes. Mars 4 and Mars 5, launched on 21 and 25 July, were intended to be orbiters but only Mars 5 succeeded. Mars 6 and Mars 7, launched on 5 and 9 August, were intended to be landers, but Mars 7 missed the red planet and communications were lost with Mars 6 before it landed.

By May 1973, the Soviet Union had launched three space station modules. Salyut 1 had been occupied only by the ill-starred Soyuz 11 crew and the second two would-be Salyut 2s had been lost.

Against this backdrop, the United States launched its own Skylab space station on 14 May. Skylab, with several times the habitable volume of the Salyut-class space stations, was launched in the last operational launch of the huge Saturn 5 vehicle that had been used in all the Apollo lunar missions. The Skylab deployment went more smoothly than either of the Soviet attempts in 1973, but was nevertheless severely marred by the loss of its meteorite shield and its left solar panel wing during launch. Thus, when the three man Apollo/Skylab 1 crew arrived on 25 May to occupy the station, they had a great deal of difficulty docking with the damaged station and it took them the better part of a day to make the minimum repairs needed to make the station habitable. Nevertheless, the crew succeeded, and they remained aboard the

Continued on page 77

Above: **This photo of Skylab was taken from the Skylab 2 mission crew's Apollo spacecraft during an inspection approach. The second Soviet lunar rover, Lunokhod 2** *(below)* **doubled its predecessor's lunar mileage.** *At right:* **Photographed during an EVA, Skylab 3 astronaut Jack Lousma's visor reflects the Earth.**

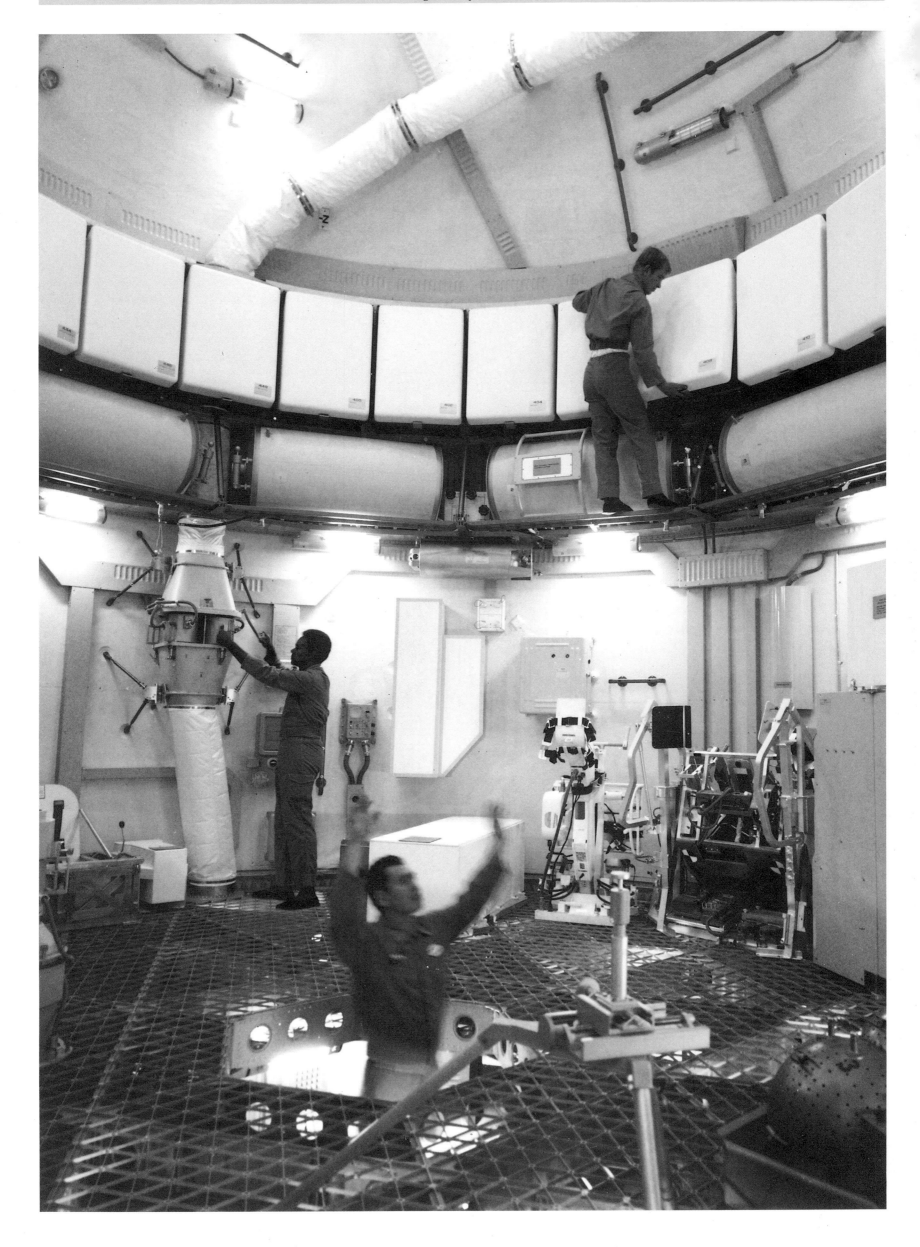

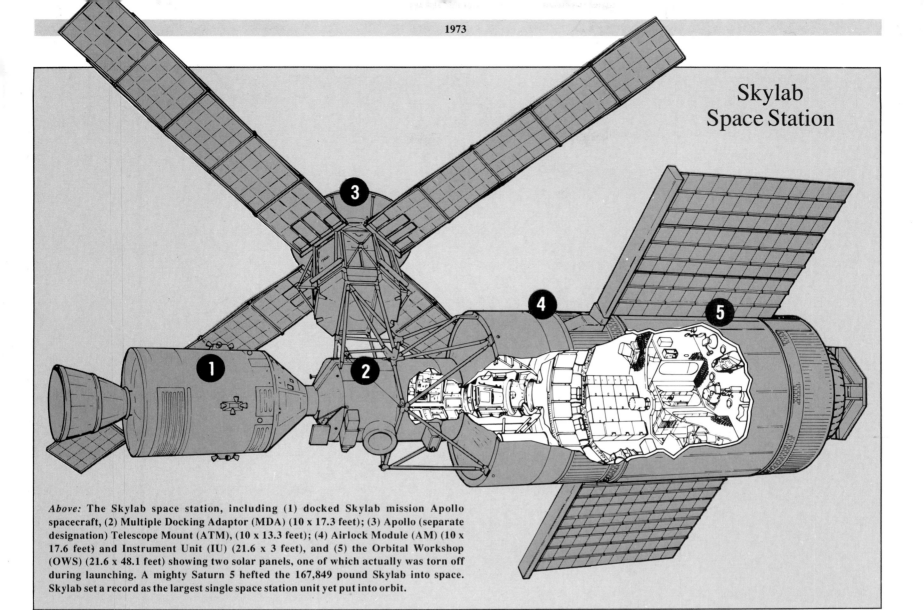

Skylab Space Station

Above: The Skylab space station, including (1) docked Skylab mission Apollo spacecraft, (2) Multiple Docking Adaptor (MDA) (10 x 17.3 feet); (3) Apollo (separate designation) Telescope Mount (ATM), (10 x 13.3 feet); (4) Airlock Module (AM) (10 x 17.6 feet) and Instrument Unit (IU) (21.6 x 3 feet), and (5) the Orbital Workshop (OWS) (21.6 x 48.1 feet) showing two solar panels, one of which actually was torn off during launching. A mighty Saturn 5 hefted the 167,849 pound Skylab into space. Skylab set a record as the largest single space station unit yet put into orbit.

Space Flight Highlights of 1973

	Launch date	Launch vehicle	Launch weight
Luna 21 (USSR)	8 Jan	D-1e	12,320 lb
Pioneer 11 (USA)	6 Apr	Atlas/Centaur	570 lb
Skylab 1 (USA)	14 May	Saturn 5	30,200 lb (CSM)
Apollo/Skylab 2 (USA)	25 May	Saturn 1B	167,849 lb
Apollo/Skylab 3 (USA)	28 Jul	Saturn 1B	30,200 lb (CSM)
Apollo/Skylab 4 (USA)	16 Nov	Saturn 1B	30,200 lb (CSM)
Soyuz 12 (USSR)	27 Sep	A-2	14,454 lb
Mariner 10 (USA)	3 Nov	Atlas/Centaur	1108 lb
Soyuz 13 (USSR)	18 Dec	A-2	14,696 lb

American Astronauts

10.	Charles Conrad (4th flight)	(Apollo/Skylab 2)	25 May
35.	Joseph Kerwin		
36.	Paul Weitz		
24.	Alan Bean (2nd flight)	(Apollo/Skylab 3)	28 Jul
37.	Owen Garriott		
38.	Jack Lousma		
39.	Gerald Carr	(Apollo/Skylab 4)	16 Nov
40.	Edward Gibson		
41.	William Pogue		

Soviet Cosmonauts

26.	Vasily Lazarev	(Soyuz 12)	27 Sep
27.	Oleg Makarov		
28.	Pyotr Klimuk	(Soyuz 13)	18 Dec
29.	Valentin Lebedev		

Opposite: The interior of Skylab's orbital workshop in construction on Earth. Above this room is the instrument unit, and below it is the crew quarters. *Above:* A Saturn 1B gets the Skylab 3 mission off the launch pad. Skylab stayed in Earth orbit from 14 May 1973 until 11 July 1979.

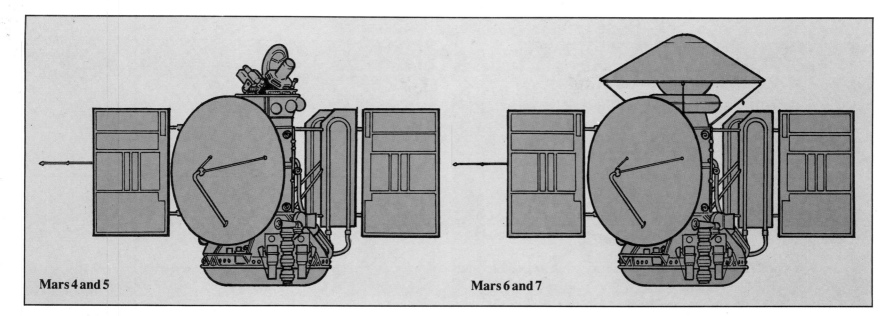

Mars 4 and 5   Mars 6 and 7

Continued from page 72

station for a record 672 hours and 50 minutes before returning to earth on 22 June. Just five weeks later, on 28 June, a second Skylab crew reached Skylab for a turn aboard the station. This time problems occurred with a fuel leak in the crew's Apollo spacecraft. A rescue mission was mounted but proved unnecessary. The crew were able to complete their docket of scientific projects and their extended stay aboard Skylab, which set a new record of 1427 hours and nine minutes.

On 27 September, two days after the second Skylab crew returned home, the Soviet Union launched the two day flight of Soyuz 12, their first manned mission since the loss of the Soyuz 11 crew more than two years before. The Soyuz spacecraft had been greatly modified, and because the cosmonauts wore bulky space suits, there was room for only a two man crew. The decision to require space suits was directly related to the Soyuz 11 disaster; had *that* crew been wearing space suits, the loss of pressure would not have affected them and they would have survived.

On November 16, the United States launched the third and last scheduled Skylab crew. This time there were no serious problems to compare with the first two missions.

On 18 December, the Soviets launched the eight-day mission of Soyuz 13, and for those eight days both astronauts and cosmonauts were in earth orbit at the same time. The three Americans—Gerald Carr, Edward Gibson and William Pogue—remained in space when 1973 ended, with more than a month to go in their sojourn aboard Skylab.

While the Soviets endured setbacks in both interplanetary exploration and space station programs during 1973, for the United States it was an auspicious year for both endeavors. Not only did the United States end 1973 with three men in space, but they also closed with three important interplanetary spacecraft plunging into the darkness of the solar system. Pioneer 10 and 11 were sister ships designed to probe the distant solar system, while Mariner 10 was designed to observe the *inner* solar system. Pioneer 11 was launched on 6 April, on the heels of Pioneer 10 which went up in 1972. Having been launched into an ideal 'window' of opportunity, Pioneer 10 passed Jupiter on 3 December 1973, just 21 months after launch, returning the first-ever closeup pictures of the Solar System's largest planet. For Pioneer 11, as well as for Mariner 10, their planetary encounters would fill the headlines of 1974.

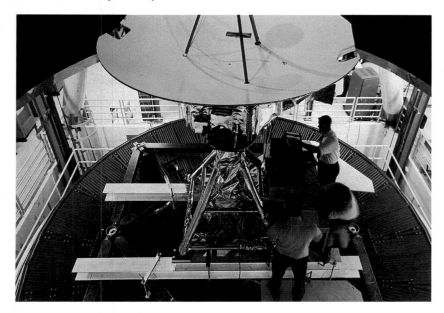

Opposite: Pioneer 11 during a pre-launch inspection. Behind it is a mockup of its Atlas/Centaur booster's third stage. *Top of page:* Mars 4/5 were intended as Mars orbiters, and Mars 6/7 were intended as landers. *At left:* Pioneer 10 during testing. *Above:* Pioneer 11 blasts off from Kennedy Space Center.

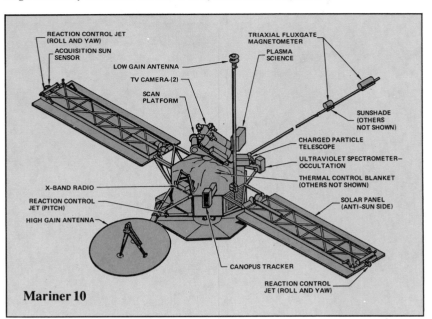

Opposite: Westar 2, launched on 10 October 1974 (foreground: Westar 3). Westar 1 was the US' first Earth-synchronous domestic communications satellite. *Below:* The Mariner 10 spacecraft. *Left:* A Mariner 10 photo of Mercury's northern limb. *Bottom of page:* The 800-mile ring basin on the left of this Mariner 10 photomosaic is the largest Mercury land feature discovered by the Mariner 10 probe.

Mariner 10

REACTION CONTROL JET (ROLL AND YAW)
ACQUISITION SUN SENSOR
LOW GAIN ANTENNA
TV CAMERA (2)
SCAN PLATFORM
X-BAND RADIO
REACTION CONTROL JET (PITCH)
HIGH GAIN ANTENNA
CANOPUS TRACKER
TRIAXIAL FLUXGATE MAGNETOMETER
PLASMA SCIENCE
SUNSHADE (OTHERS NOT SHOWN)
CHARGED PARTICLE TELESCOPE
ULTRAVIOLET SPECTROMETER—OCCULTATION
THERMAL CONTROL BLANKET (OTHERS NOT SHOWN)
SOLAR PANEL (ANTI-SUN SIDE)
REACTION CONTROL JET (ROLL AND YAW)

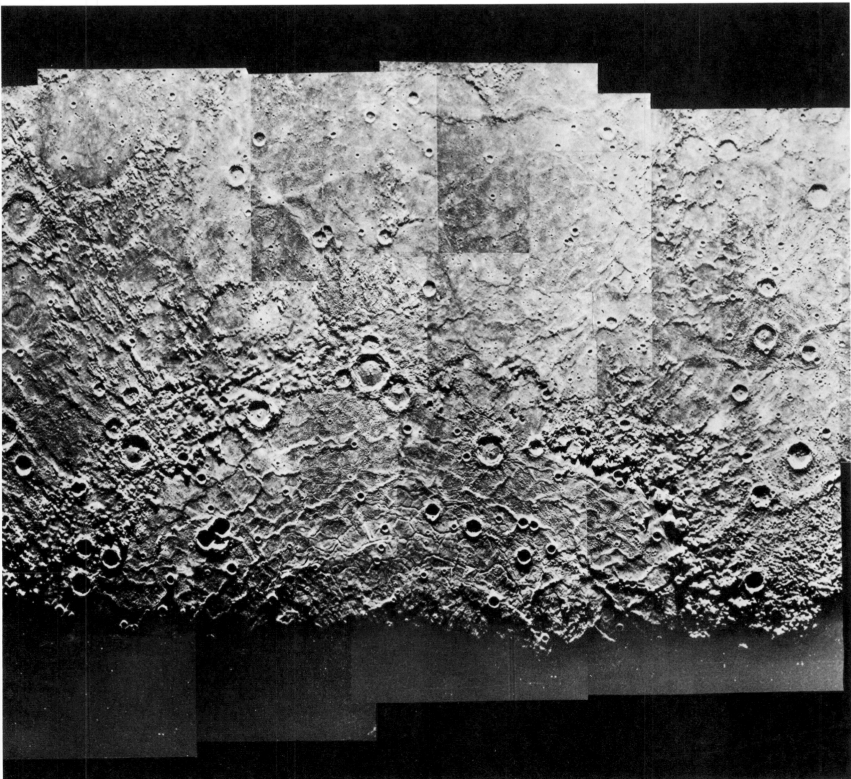

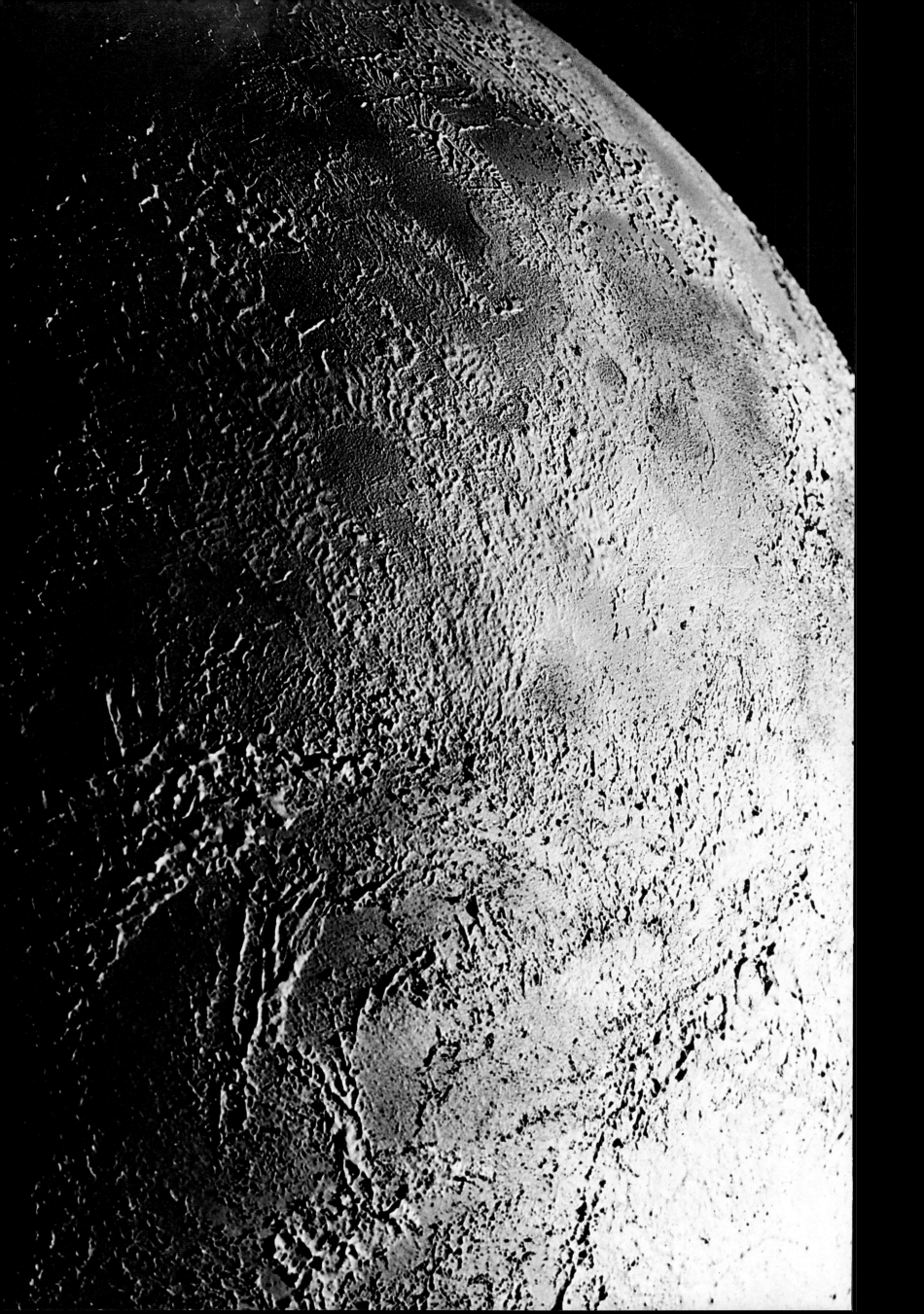

These pages: In this artist's conception, Mariner 10 is portrayed as it encountered the planet Mercury. Launched atop an Atlas/Centaur on 3 November 1973, Mariner 10 was designed to conduct investigations of both Venus and Mercury, including environmental, atmospheric, geological and planetary characteristics of Mercury, and the structure, composition, circulation and distribution characteristics of Venus' extremely dense atmosphere. Mariner 10 passed Venus on 5 February 1974, and encountered Mercury for the first time on on 29 March 1974. Mariner 10 flew by Mercury again—a day earlier than was planned—on 21 September 1974. The 1100-pound planetary probe flew by Mercury for an unplanned third time in 1975. Mariner 10 transmitted some 8000 photographs altogether, and was shut down and placed in permanent solar orbit on 24 March 1975.

1975

Having placed Salyut 4 into orbit in the last days of 1974, the Soviet Union launched Soyuz 17 with the new station's first visitors on 10 January. The first crew completed their tenure aboard Salyut 4 on 9 February, but Soyuz 18, launched on 5 April, failed to achieve orbit and was forced to make an emergency landing in western Siberia. A substitute for this mission, designated Soyuz 18B, was launched on 24 May.

The high point of the manned space flight efforts of both the United States and the Soviet Union came in July with the Apollo-Soyuz Test Project (ASTP) in which the last ever American Apollo mission docked with the Soviet Soyuz 19. The long-rehearsed exercise began on 15 July when the two spacecraft were launched from the Kennedy and Baikonur space centers, respectively. The docking took place on 17 July and the hatch opened to permit the first international handshake in space. On the American side were Donald 'Deke' Slayton, the only member of the original group of Mercury astronauts still on active duty, and the only Mercury astronaut who had previously never flown in space. Slayton's companions were space rookie Vance Brand and Tom Stafford, veteran of Gemini 6, Gemini 9 and Apollo 10.

On the Soviet side were Alexei Leonov, veteran of Voskhod 2 and soon to be named head of the cosmonaut corps, and Valeri Kubasov, a veteran of Soyuz 6. The two ASTP spacecraft remained together for two days, practicing a redocking exercise on 19 July. Soyuz 19 returned to earth on 21 July and America's last Apollo mission ended three days later. On 26 July, the crew of Soyuz 18B, who had been in space during the entire ASTP exercise, returned to earth after more than two months in space. Soyuz 20, launched on 17 November, was an unmanned 'supply ship' that docked with Salyut 4 by remote control.

On 20 August and 9 September, the United States launched Viking 1 and Viking 2. Bound for Mars, they were the first American attempt to soft land a spacecraft on another planet. The Vikings were to be a major milestone in the American space program when they touched down in 1976.

Above: **Thomas Stafford and Alexei Leonov in the hatchway connecting Apollo and Soyuz during the Apollo-Soyuz Test Project.** *Below:* **Stafford, Leonov and Slayton.** *At right:* **An artist's concept of the Apollo (near)-Soyuz (far) docking maneuver.**

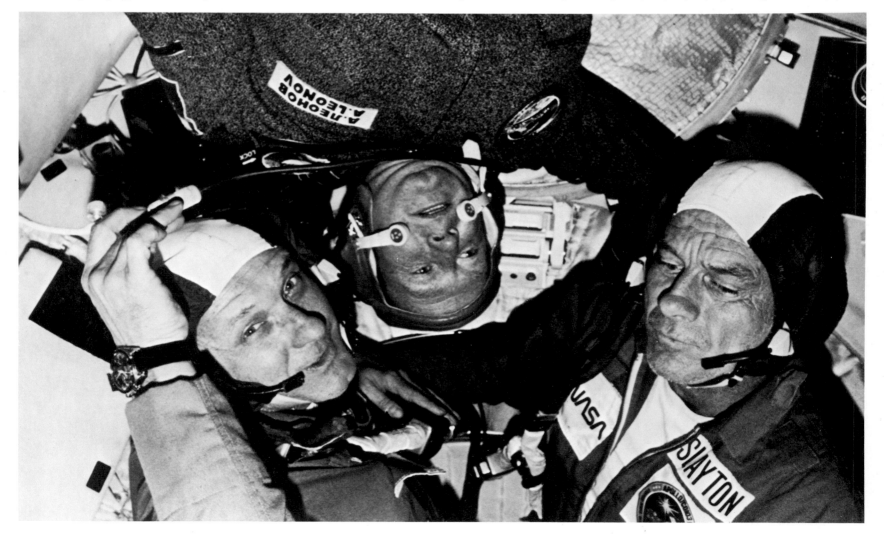

At left: Soyuz 19, photographed from the rendezvous window of the Apollo spacecraft. Apollo, being more maneuverable, was the active partner in the ASTP docking. *Above:* Apollo docking module pilot Donald K 'Deke' Slayton and Soyuz commander Alexei Leonov yuck it up in an impromptu Apollo-Soyuz Test Project onboard activity. *Below:* Soyuz 19 and Apollo in docking approach and *below left*, two views of the Soyuz 19 docking module.

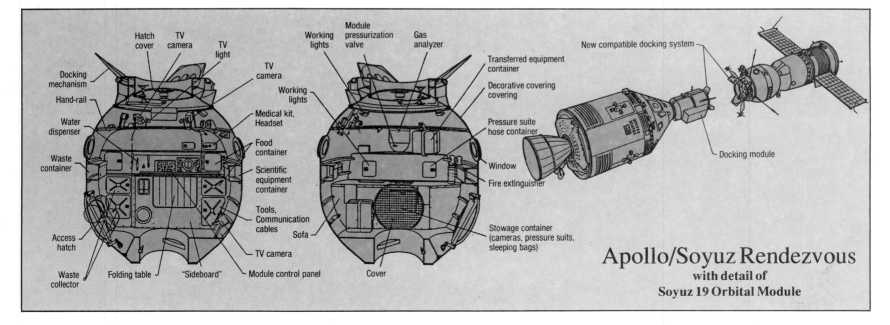

Docking mechanism
Hand-rail
Water dispenser
Waste container
Access hatch
Waste collector
Folding table
"Sideboard"
Hatch cover
TV camera
TV light
TV camera
Working lights
Medical kit, Headset
Food container
Scientific equipment container
Tools, Communication cables
Sofa
TV camera
Module control panel
Working lights
Module pressurization valve
Gas analyzer
Cover
Transferred equipment container
Decorative covering covering
Pressure suit hose container
Window
Fire extinguisher
Stowage container (cameras, pressure suits, sleeping bags)
New compatible docking system
Docking module

Apollo/Soyuz Rendezvous
with detail of
Soyuz 19 Orbital Module

Top of page: The Viking 1 Mars probe is shown here receiving its launch shroud. *Above:* The two parts of Viking 2 are here being mated—the encapsuled Lander and the boxy Orbiter. *At right:* Viking 2 heads for Mars on 19 September 1975, from Launch Pad 41 at NASA's Kennedy Space Center, at Cape Canaveral, Florida.

The Intelsat 4 design *(at left)* was represented by eight satellites launched from 1971–1975. The six satellites representing the Intelsat 4-A design *(above* and *below)* were launched throughout the period of 1975–1978 and were, due to improvements in their direction-intensive spot-beam antennas, more powerful than the Intelsat 4s. *At right:* Landsat 2 launch. See Landsat data on page 205.

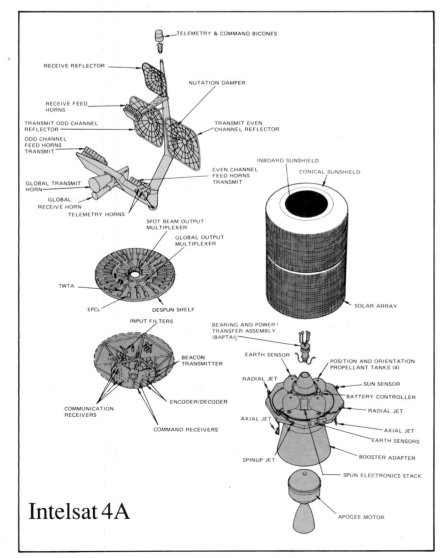

Intelsat 4A

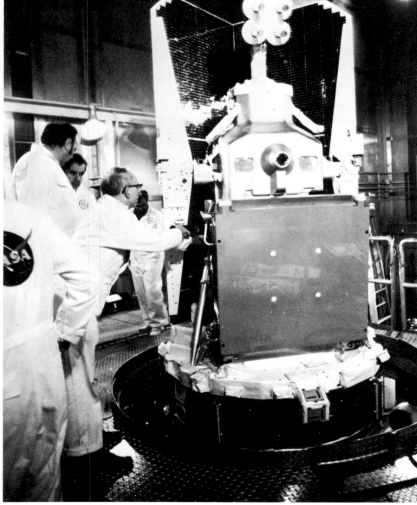

Space Flight Highlights of 1975

	Launch date	Launch vehicle	Launch weight
Soyuz 17 (USSR)	11 Jan	A-2	14,454 lb
Landsat-2 (USA)	22 Jan	Delta	1965 lb
GEOS-C (USA)	9 Apr	Delta	750 lb
Intelsat 4 (USA)*	22 May	Atlas/Centaur	1600 lb
Soyuz 18B (USSR)	24 May	A-2	14,454 lb
Venera 9 (USSR)	8 June	D-1e	10,859 lb
Nimbus-F (USA)	12 June	Delta	1823 lb
OSO-8 (USA)	21 June	Delta	2346 lb
Soyuz 19/ASTP (USSR)	15 Jul	A-2	14,960 lb
Apollo ASTP (USA)	15 Jul	Saturn 1B	30,781 lb (CSM)
Viking 1 (USA)	20 Aug	Titan 3/Centaur	5125 lb
Viking 2 (USA)	9 Sep	Titan 3/Centaur	5125 lb
Intelsat 4A (USA)**	25 Sept	Atlas/Centaur	1600 lb
GOES-A (SMS-C) (USA)	16 Oct	Delta	1375 lb

*Last of eight Intelsat 4 launches
**First of six Intelsat 4A launches

American Astronauts

14.	Thomas Stafford (4th flight)	(ASTP)	15 Jul
42.	Vance Brand		
43.	Donald Slayton		

Soviet Cosmonauts

33.	Alexei Gubarev	(Soyuz 17)	11 Jan
34.	Georgi Grechko		
26.	Vasily Lazarev (2nd flight)	(Soyuz 18)	5 Apr
27.	Oleg Makarov (2nd flight)		
28.	Pyotr Klimuk (2nd flight)	(Soyuz 18B)	24 May
22.	Vitali Sevastyanov (2nd flight)		
11.	Alexei Leonov (2nd flight)	(Soyuz 19 ASTP)	15 Jul
18.	Valeri Kubasov (2nd flight)		
	Unmanned	(Soyuz 20)	17 Nov

At left: Landsat 2, shown in preparation for its 22 January 1975 launch date, teamed up with Landsat 1, launched in 1972, to provide Earth resources information on nearly the entire globe. *Top of page:* The Ford GOES-A, another chapter in NASA's 24-hour Synchronous Weather Satellite launch history. Powerful Nimbus-F *(above)* could process up to eight incoming signals at a time. *At right:* The Venera 9 and 10 Orbiter/Lander configuration. Venera 9 provided the first television pictures of Venus' surface.

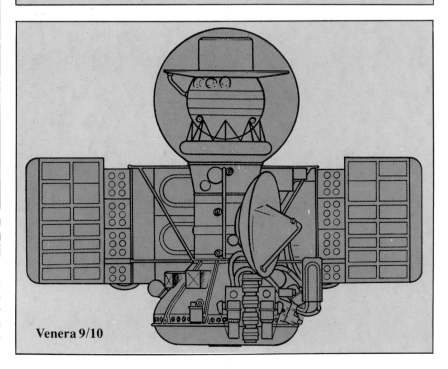

Venera 9/10

Above: GEOS-C measured ocean tides, sea state, gravity and solid-earth dynamics, and also participated in a satellite-satellite tracking experiment. This was, in addition, the first independent GEOS satellite launch, as GEOS-A and -B had been included with Explorers 29 and 36, respectively. *At right:* A Hughes Aircraft Company technician checks one of OSO-8's two spectrometers, which took high resolution ultraviolet measurements. OSO-8, the last OSO (Orbiting Solar Observatory) satellite, performed solar physics studies and took readings on X-ray sources in the Milky Way and beyond. The OSO program provided continuous data on the Sun's 11-year sunspot cycle.

1976

The United States' bicentennial year was marked by a series of payoffs in that country's planetary exploration efforts. On 10 February, Pioneer 10 reached Saturn after a flight of almost four years. The photos that the spacecraft transmitted were the closest views yet of the great ringed planet.

On 19 June, Viking 1 was inserted into orbit around Mars and on 20 July the Viking 1's lander successfully touched down on the Martian surface, transmitting the first photographs of the mysterious planet ever to have been taken at eye level. The Viking 2 orbiter joined its sister ship in Mars orbit on 7 August and its lander also successfully descended to Mars' dusty surface. The Viking 2 lander remained operational until it was powered down on 10 November, but the Viking 1 lander was kept running for four years, observing the seasonal cycles on Mars and the gentle comings and goings of Martian snowstorms.

Soviet space activity revolved around operation of the Salyut 5 space station, which was launched on 22 June. Soyuz 21, the only crew to utilize Salyut during 1976, was launched on 6 July and returned on 9 August. The Soyuz 22 mission, launched on 15 September and returning on 23 September, was unrelated to Salyut 5 and is thought by some western observers to have been a reconnaissance mission directed at obtaining certain specific high resolution photographs.

Soyuz 23, on the other hand, *was* intended to dock with Salyut. However, a failure in the craft's automatic docking system resulted in an abort for the 14 October launch mission.

Another milestone of planetary exploration that was overshadowed by the Vikings was the mission of the Luna 24 probe. Launched on 18 August, between the arrival of the two Viking landers, Luna 24 was the third successful Soviet sample return mission to the moon. The most complex of that class of lander, the spacecraft succeeded in drilling into the moon's surface and returning with a six-foot core sample from the Sea of Crises. The true milestone of Luna 24 was, however, that it marked the end of 17 years of continuous lunar exploration that had begun in 1959 when Luna 3 photographed the dark side of the moon for the first time. In the intervening years the Soviet Union had conducted seven unmanned soft landings on the moon, while the United States had conducted four successful unmanned and six successful manned landings. Since 1976, neither superpower has conducted a lunar exploration of any kind.

Above: Satcom 2 rises from Kennedy Space Center's Launch Complex 17. Satcom 2 *(below)* was built and put into service by RCA, and was launched for RCA by NASA. *At right:* Comstar satellites were built by Hughes Aircraft principally for AT&T and GTE Telecommunications, and handled telephone service for the US and Puerto Rico.

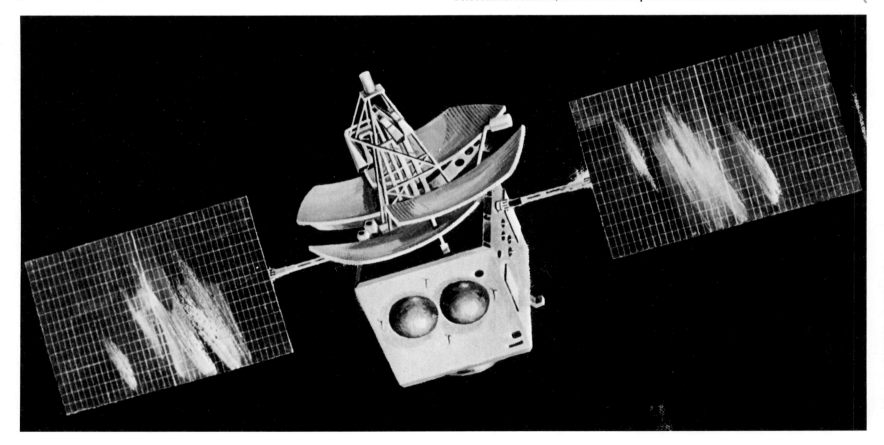

Space Flight Highlights of 1976

	Launch date	Launch vehicle	Launch weight
Marisat A (USA)	19 Feb	Delta	1445 lb
RCA Satcom 2 (USA)	26 Mar	Delta	1900 lb
LAGEOS (USA)	4 May	Delta	903 lb
Comstar 1 (USA)	13 May	Atlas/Centaur	3300 lb
Salyut 5 (USSR)	22 Jun	D-1	40,920 lb
Soyuz 21 (USSR)	6 Jul	A-2	14,454 lb
Soyuz 22 (USSR)	15 Sep	A-2	14,300 lb
Soyuz 23 (USSR)	14 Oct	A-2	14,454 lb

American Astronauts

(No Americans would go into space in the period from 1976 through 1980. A planned continuation of Apollo was cancelled and the Space Shuttle was not available until 1981.)

Soviet Cosmonauts

35.	Boris Volynov	(Soyuz 21)	6 Jul
36.	Vitaly Zholobov		
5.	Valery Bykovsky (2nd flight)	(Soyuz 22)	15 Sep
37.	Vladimir Aksyonov		
38.	Vyacheslav Zudov	(Soyuz 23)	14 Oct
39.	Valery Rozhdestvensky		

At left: The LAGEOS (Laser Geodynamics Satellite) was a solid sphere which reflected groundbased laser pulses in an experiment to measure the motion of the Earth's crust. *Above:* The first Marisat maritime communications satellite was one of three parts of a Comsat Corp system. *Right:* Another Comsat venture, Comstar 1. Compare with Intelsat 1 on page 33.

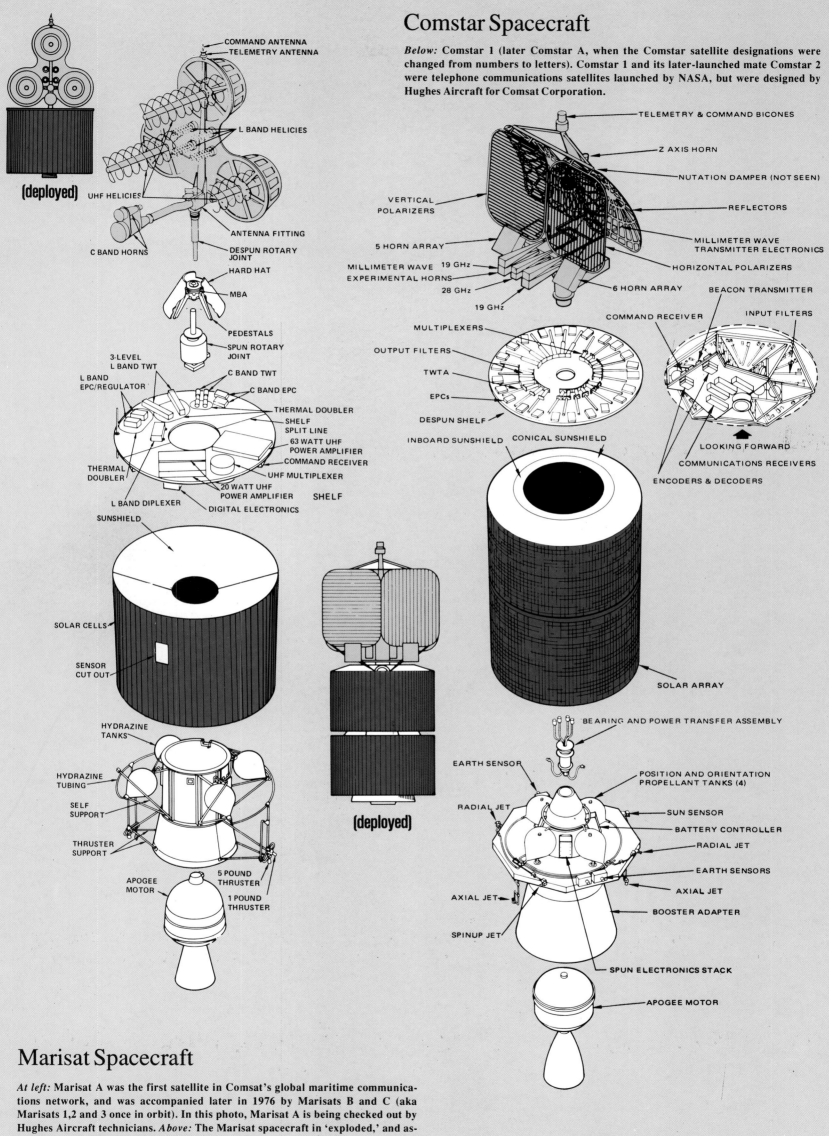

Comstar Spacecraft

Below: Comstar 1 (later Comstar A, when the Comstar satellite designations were changed from numbers to letters). Comstar 1 and its later-launched mate Comstar 2 were telephone communications satellites launched by NASA, but were designed by Hughes Aircraft for Comsat Corporation.

(deployed)

COMMAND ANTENNA
TELEMETRY ANTENNA
L BAND HELICIES
UHF HELICIES
C BAND HORNS
ANTENNA FITTING
DESPUN ROTARY JOINT
HARD HAT
MBA
PEDESTALS
SPUN ROTARY JOINT
3-LEVEL L BAND TWT
L BAND EPC/REGULATOR
C BAND TWT
C BAND EPC
THERMAL DOUBLER
SHELF SPLIT LINE
63 WATT UHF POWER AMPLIFIER
COMMAND RECEIVER
UHF MULTIPLEXER
THERMAL DOUBLER
20 WATT UHF POWER AMPLIFIER
L BAND DIPLEXER
DIGITAL ELECTRONICS
SHELF
SUNSHIELD
SOLAR CELLS
SENSOR CUT OUT
HYDRAZINE TANKS
HYDRAZINE TUBING
SELF SUPPORT
THRUSTER SUPPORT
APOGEE MOTOR
5 POUND THRUSTER
1 POUND THRUSTER

(deployed)

TELEMETRY & COMMAND BICONES
Z AXIS HORN
NUTATION DAMPER (NOT SEEN)
REFLECTORS
MILLIMETER WAVE TRANSMITTER ELECTRONICS
HORIZONTAL POLARIZERS
6 HORN ARRAY
VERTICAL POLARIZERS
5 HORN ARRAY
MILLIMETER WAVE EXPERIMENTAL HORNS
19 GHz
28 GHz
19 GHz
MULTIPLEXERS
OUTPUT FILTERS
TWTA
EPCs
DESPUN SHELF
INBOARD SUNSHIELD
CONICAL SUNSHIELD
BEACON TRANSMITTER
INPUT FILTERS
COMMAND RECEIVER
LOOKING FORWARD
COMMUNICATIONS RECEIVERS
ENCODERS & DECODERS
SOLAR ARRAY
BEARING AND POWER TRANSFER ASSEMBLY
EARTH SENSOR
RADIAL JET
AXIAL JET
SPINUP JET
POSITION AND ORIENTATION PROPELLANT TANKS (4)
SUN SENSOR
BATTERY CONTROLLER
RADIAL JET
EARTH SENSORS
AXIAL JET
BOOSTER ADAPTER
SPUN ELECTRONICS STACK
APOGEE MOTOR

Marisat Spacecraft

At left: Marisat A was the first satellite in Comsat's global maritime communications network, and was accompanied later in 1976 by Marisats B and C (aka Marisats 1, 2 and 3 once in orbit). In this photo, Marisat A is being checked out by Hughes Aircraft technicians. *Above:* The Marisat spacecraft in 'exploded,' and assembled, views. The three Marisats were identical.

The inset *above* is a Viking Lander 1 photo. 'Big Joe,' the large boulder at photo left is approximately 6.5 feet across, and is approximately 25 feet from the lander. The dust drifts here are evidence of wind erosion.

At left: Viking Orbiter 1 took the photos of which this photomosaic is composed. The area in view here is the northeast margin of Mars' Tharsis Ridge, the planet's youngest volcanic region. Crater chains, channels and lava flows can be seen on the flanks of the vocanoes here, which range from 40–250 miles across. The prominent volcano with the channel running from its summit crater to a crater at its base is known as Ceraunius Tholus, of which it is postulated that the base crater is that of a huge meteorite impact that triggered this mountain's most recent volcanic eruption. The largest fault in the fault zone at photo left is called Ceraunius Fossae, and the volcano just above Ceraunius Tholus is called Uranius Tholus.

The very evident meteorite craters here are surrounded by 'coronas' of ejectamenta—Martian surface material which was displaced by the various meteorite impacts. Viking 1 captured these images on 15 November 1977.

Below: 'On a clear day on Mars, you can see tens of thousands of rocks.' So NASA's photocaptioning department says, and this Viking Lander 2 photomosaic certainly seems to bear out that statement. The large rock at photo center is approximately two feet long and one foot high. Winding from upper left to lower right in front of the aforesaid rock is a feature that appears to be a small channel.

The horizon here is actually nearer level than it appears, for the spacecraft itself was tilted at an eight degree angle. As discussed elsewhere in this volume's captions, the rocks seen here are probably limonite, familiar on Earth as iron ore. Mars' ruddy complexion has to do with the planet's iron-rich surface materials, which have turned 'rust-red' as they have been oxidized by the water vapor which is present in Mars' predominately carbon dioxide atmosphere.

The idea of life on Mars has long kindled the human imagination, and with the increase in man's scientific knowledge about 'the Red Planet,' the notion has taken on some plausibility. Viking Lander soil tests for bacteria proved inconclusive, though there are some indications that Mars once had a thicker atmosphere, and rivers.

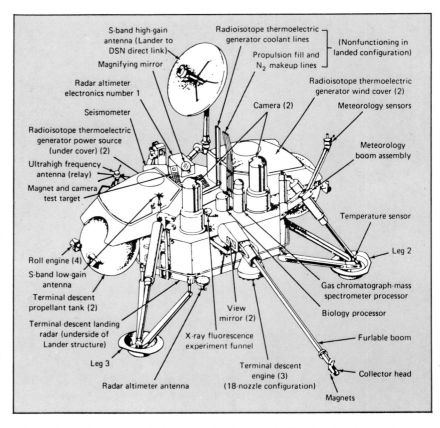

Above: The Viking Lander design. Each Lander was sterilized and sealed inside a capsule (see also page 88) which was jettisoned upon the Lander's descent to Mars; this was to avoid contaminating Mars with Earth-born microbes. The capsule's disc-like shape belied the Earthling myth that all flying saucers *come from* Mars.

Above: Soon after Viking 1 landed, the protective cover for the craft's soil sample scoop was ejected, and its first bounce disturbed some soil, as indicated here. At photo right is the footpad of one of the Viking Lander's legs.

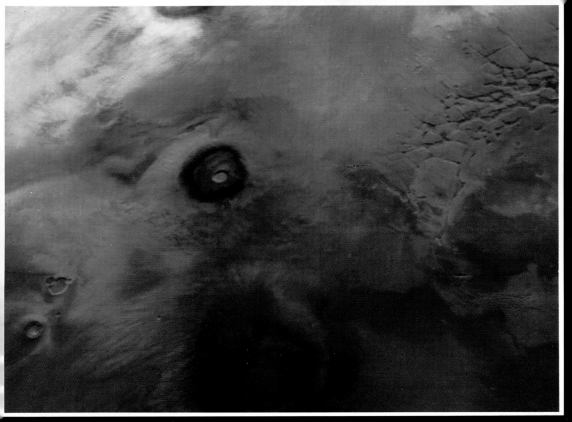

At left: This oblique view of the Martian area known as Tharsis Ridge was obtained by the Viking Orbiter 1, which inserted itself into a Mars orbit on 19 June 1976. Tharsis Ridge is 6.2 miles high. From the top down in this photo are the large Martian volcanoes Ascraeus Mons, Pavonis Mons and Arsia Mons, and to their left are the small volcanoes Biblis Patera and Ulysses Patera. The region of dissecting faults at upper photo right is known as Noctis Labyrinthus.

Below: The Viking Orbiter 1's partner in investigation, the Viking Lander 1, took this southeast view from its landing position on Mars' Chryse Planitia, on 21 July 1976. The rocks seen here are probably limonite—a natural iron oxide which on Earth is used as iron ore. The time of this photo is near high noon, Mars time. The sky's orange cast could well be due to dust particles suspended in Mars' lower atmosphere.

At right: A Viking Orbiter 1 view, revealing Mars' thin carbon dioxide atmospheric envelope. The large crater basins extending from the foreground into the distance were formed by collisions with large spaceborne bodies such as asteroids; the smaller craters here in evidence are meteor craters and a few volcanoes. The very large crater basin in the foreground is known as Argyre Planitia. If Mars indeed once had a thicker atmosphere (see previous very long caption), tests on these meteor craters should prove many of them to of comparatively recent origin.

1977

Having abandoned the Apollo program, the United States entered a period of five years when, for the first time since 1961, it had no ability to conduct manned space flight. However, the road back to a manned space flight capability for the United States began with a series of five successful gliding test flights of *Enterprise*, the 75-ton prototype for a series of reusable manned space planes that it was hoped would become operational in 1979, but which actually did not go into space until 1981.

The Soviet Union, meanwhile, launched four manned missions and another space station. The first of the manned missions, Soyuz 24, was on 7 February and it remained docked with Salyut 5 until 25 February. This mission was followed by the launch of the Salyut 6 space station on 29 September. The first crew intended for Salyut 6 was launched aboard Soyuz 25 on 9 October but they failed to dock with the station and had to return to earth two days later.

On 10 December, Soyuz 26 took Georgi Grechko and Yuri Romanenko to a successful docking with Salyut 6 and, as the year ended, they remained aboard with more than two months left in what would be a record-breaking stay in space.

While the United States may have temporarily abandoned its manned space flight capability, 1977 marked the launch of the two most ambitious probes of the outer solar system that are likely to be launched in this century. Voyager 1 and Voyager 2, launched on 5 September and 20 August respectively, would come much closer to Jupiter and Saturn than the Pioneers, and Voyager 2 would be the first spacecraft designed to fly past Uranus and Neptune.

In the illustration *below*, note the Salyut 6 space station's three solar panels and docking ports fore and aft. The additional docking port—a striking advance over earlier Salyuts—meant that the craft could be visited by more than one crew while already occupied, and could be supplied with less maneuvering by the Soviets' unmanned robot supply 'tugs,' which are also known as 'Progress' supply vehicles.

At right: Voyagers 1 and 2 were identical. Besides spectacular Jupiter and Saturn investigations, Voyager 2 is the first spacecraft to have photographed Uranus at close range, and is even now on its way to Neptune. *Overleaf:* Each Voyager carries a gold plaque (seen clearly on spacecraft's hull) by which friendly aliens may locate Earth—as do the Pioneers 10 and 11, which have preceded the Voyagers out of the Solar System. The Voyagers' nuclear generators will enable them to transmit data until 2007 AD.

Space Flight Highlights of 1977

	Launch date	Launch vehicle	Launch weight
Soyuz 24 (USSR)	7 Feb	A-2	14,454 lb
Palapa 2 (Indonesia)	8 Jul	Delta	674 lb
Voyager 2 (USA)	20 Aug	Titan 3/Centaur	1797 lb
Voyager 1 (USA)	5 Sep	Titan 3/Centaur	1797 lb
GMS (Japan)	14 Jul	Delta	694.4 lb
Salyut 6 (USSR)	29 Sep	D-1	41,580 lb
Soyuz 25 (USSR)	9 Oct	A-2	14,454 lb
Soyuz 26 (USSR)	10 Dec	A-2	14,454 lb

American Astronauts

No Americans went into space during 1977, but astronauts Joe Eagle, Gordon Fullerton, Fred Haise and Richard Truly participated as two-man teams in five glide tests of a Space Shuttle Orbiter between 12 August and 26 October.

Soviet Cosmonauts

21.	Victor Gorbatko (2nd flight)	(Soyuz 24)	7 Feb
40.	Yury Glazkov		
41.	Vladimir Kovalenok	(Soyuz 25)	9 Oct
42.	Valery Ryumin		
43.	Yuri Romanenko	(Soyuz 26)	10 Dec
34.	Georgi Grechko (2nd flight)		

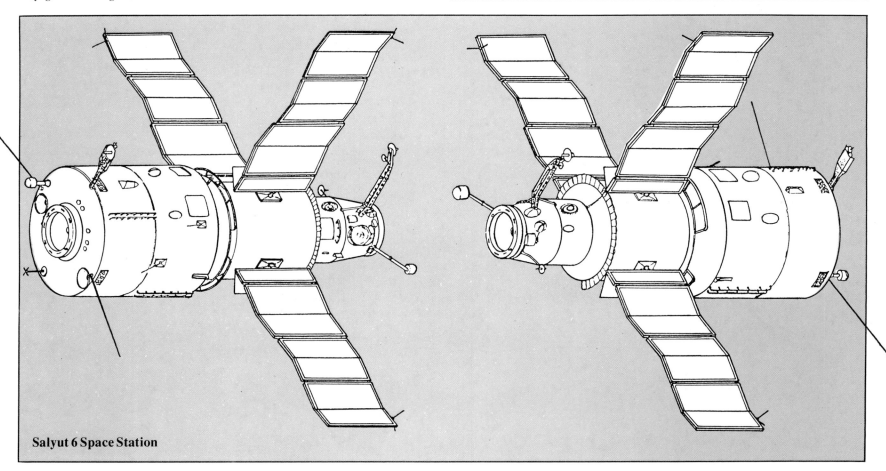

Salyut 6 Space Station

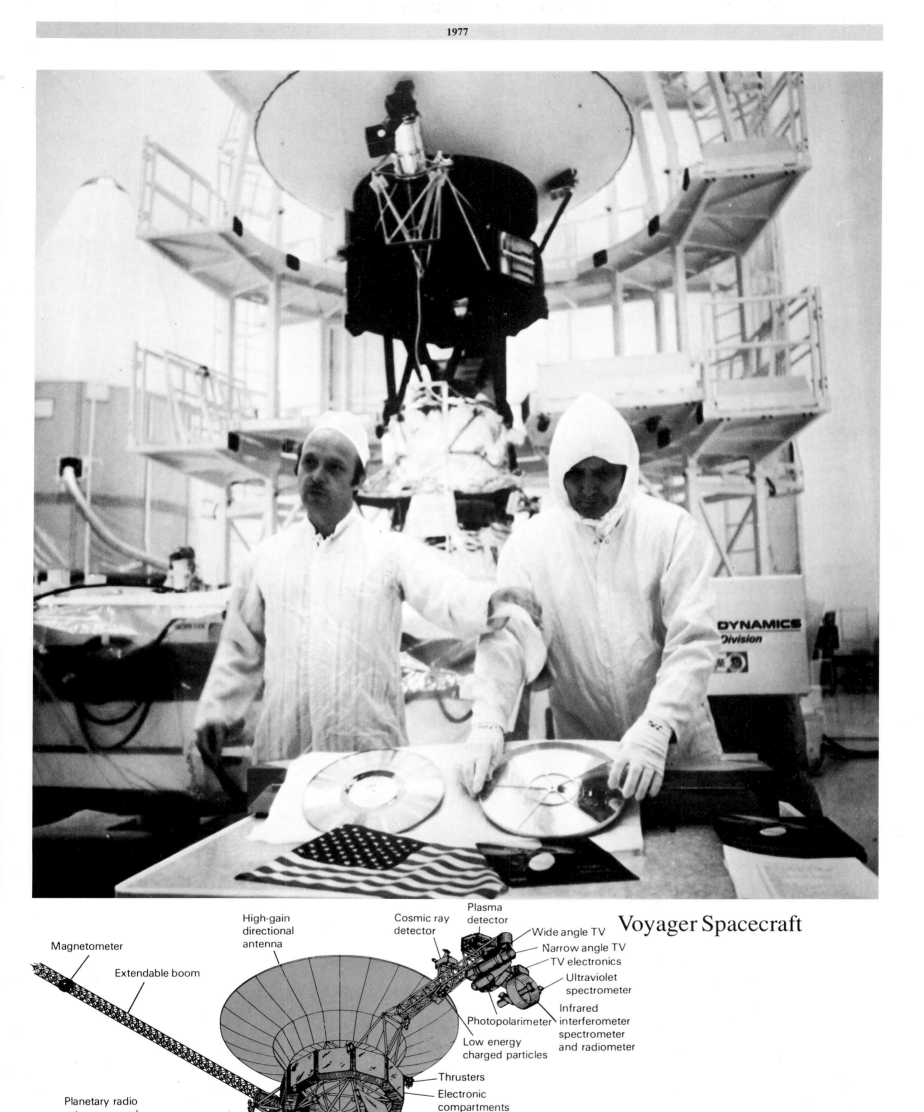

Voyager Spacecraft

Magnetometer

Extendable boom

High-gain directional antenna

Cosmic ray detector

Plasma detector

Wide angle TV

Narrow angle TV

TV electronics

Ultraviolet spectrometer

Photopolarimeter

Infrared interferometer spectrometer and radiometer

Low energy charged particles

Planetary radio astronomy and plasma wave antenna

Thrusters

Electronic compartments

Science instrument calibration panel and shunt radiator

Propulsion fuel tank

Radioisotope thermoelectric generators

Planetary radio astronomy and plasma wave antenna

Opposite: The Palapa 2 telecommunications satellite in testing at the Hughes Aircraft Company, who built it for Indonesia. *Above:* The Voyager 1 and 2 configuration. *Top of page:* Besides instruments and the plaque discussed on page 106, the Voyagers each carry an American flag and a 'Sounds of Earth' recording—with greetings in 60 languages, music, and animal, ocean and wind sounds. Here, the flag and record are being readied for Voyager 2.

111

Above: The *Enterprise* Orbital Vehicle (OV-101) was not actually designed for space, but rather as a glider prototype for similar craft which would truly go into orbit. Here, *Enterprise* is being mounted upon its Boeing 747 launch plane for a glide test. *Below:*

Enterprise shows its landing capability with wheels down, just like a regular airplane! *At right:* Japan's Geostationary Meteorological Satellite, built by Hughes Aircraft. Japan has had an active space program since the 1950s.

1978

Soyuz 26 Cosmonauts Georgi Grechko and Yuri Romanenko were aboard the Salyut 6 space station when the new year began, and on 11 January the crew of Soyuz 27 (launched 10 January) arrived for a five-day visit after which they went home aboard Soyuz 26, leaving the fresher Soyuz 27 spacecraft docked with Salyut. On 22 January Grechko and Romanenko took delivery of supplies from the Progress 1 robot transport that was launched on 20 January and remotely directed to a docking with Salyut 6. On 3 March the Soyuz 26 cosmonauts received a second group of guests who had come to visit them via Soyuz 28. This time the crew was Soviet cosmonaut Alexei Gubarov and Vladimir Remek, a *Czechoslovakian* cosmonaut and the first man in space who was a citizen of neither the United States or the Soviet Union.

When Grechko and Romanenko finally returned to earth aboard Soyuz 27 on 16 March, the list of first time accomplishments of their tenure aboard Salyut was impressive: (1) two crews had used a space station simultaneously on *two* occasions; (2) a remote-controlled supply ship had supplied a crew in space; (3) two crews had exchanged spacecraft in space; (4) a person from a third country had flown in space for the first time; and last, but certainly not least (5) Grechko and Romanenko had set a new space endurance record of 96 days and 10 hours, bettering the record of the final American Skylab by more than *twelve* days!

Opposite: **Cosmonauts Vladimir Kovalenok and Alexander Ivanchenkov docked their Soyuz 29 spacecraft with Salyut 6 and spent a record 139 and 15 hours days in space.** *Below:* **Landsat 3, in pre-launch testing at General Electric's Valley Forge, Pennsylvania facility.** *Below right:* **RCA technicians inspect TIROS-N, the last of the long, successful TIROS weather satellite line.**

What was even more impressive about Soviet manned space flight accomplishments for 1978 was that, having accomplished what they did with Soyuz 26 / 27 / 28, they did it again!

On 15 June Vladimir Kovalenok and Alexander Ivanchenkov were launched up to Salyut 6 aboard Soyuz 29. On 27 June they were joined by Soviet cosmonaut Pyotr Klimuk and Polish cosmonaut Miroslaw Hermaszewski aboard Soyuz 30. These visitors departed on 5 July to make way for the Progress 2 tug on 7 July, which was followed by Progress 3 on 8 August. On 27 August, Soviet cosmonaut Valery Bykovsky and East German cosmonaut Sigmund Jaehn joined Kovalenok and Ivanchenkov aboard the space station. Having arrived in Soyuz 31, these visitors went home in Soyuz 29, again leaving a fresh spacecraft for the original crew. The arrival of Progress 4 in October was almost anticlimactic.

When Kovalenok and Ivanchenkov finally came home aboard Soyuz 31 on 2 November they brought with them the second new space endurance record in 1978—139 days and 15 hours!

The major space effort by the United States during 1978 was the launching of the unmanned Pioneer Venus 1 on 20 May and Pioneer Venus 2 on 8 August. These spacecraft arrived at Venus on 4 and 9 December respectively. Pioneer Venus 1 was placed into orbit around the cloudy planet, while its companion released five scientific probes which descended into the planet's dense atmosphere, transmitting data about the thick, swirling carbon dioxide clouds that surged with sulfuric acid droplets and crackled with lightning. While none of the probes were intended to survive all the way to the planet's surface, one did, and it survived the intense pressure and 400 ° F heat for over an hour.

Above: Technicians prepare the European Space Agency's second GEOS/ESA satellite for NASA launch. *Opposite page*, clockwise from above left: The Czechoslovakian Intercosmonaut Vladimir Remek gained international fame as the first man in space from a country other than the US or the USSR; and, also a part of the Soviet Intercosmos program, East Germany's Sigmund Jähn flew on board Soyuz 31; this row of Soyuz mission patches includes—from left to right—the patch usually worn by Soviet cosmonauts, an Intercosmos patch for cooperative missions with foreign countries, an Intercosmos patch for the Czech mission, an Intercosmos patch for the Polish mission, and an Intercosmos patch for the East German mission; and this photo is of Intercosmonaut Miroslaw Hermaszewski of Poland and (left) Soviet cosmonaut Pyotr Klimuk.

Pioneer Venus Orbiter

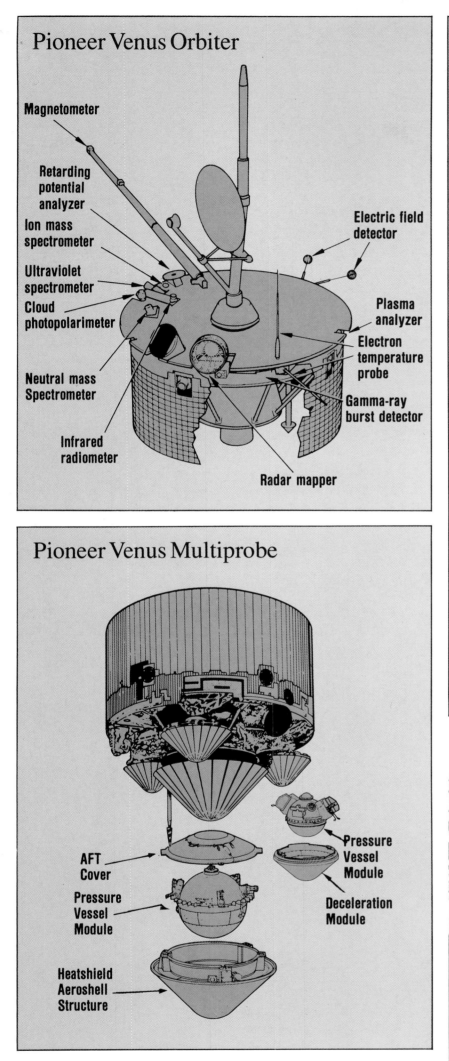

Magnetometer

Retarding potential analyzer

Ion mass spectrometer

Ultraviolet spectrometer

Cloud photopolarimeter

Neutral mass Spectrometer

Infrared radiometer

Electric field detector

Plasma analyzer

Electron temperature probe

Gamma-ray burst detector

Radar mapper

Pioneer Venus Multiprobe

AFT Cover

Pressure Vessel Module

Heatshield Aeroshell Structure

Pressure Vessel Module

Deceleration Module

Space Flight Highlights of 1978

	Launch date	Launch vehicle	Launch weight
Soyuz 27 (USSR)	10 Jan	A-2	14,454 lb
Soyuz 28 (USSR)	2 Mar	A-2	14,454 lb
Landsat 3 (USA)	5 Mar	Delta	2163 lb
Pioneer Venus 1 (USA)	20 May	Atlas/Centaur	1283 lb
Soyuz 29 (USSR)	15 Jun	A-2	14,454 lb
GOES-C (USA)	16 Jun	Delta	1375 lb
Soyuz 30 (USSR)	27 Jun	A-2	14,454 lb
GEOS (ESA)	14 Jul	Delta	1260 lb
Pioneer Venus 2 (USA)	8 Aug	Atlas/Centaur	1993 lb
Soyuz 31 (USSR)	26 Aug	A-2	14,454 lb
TIROS-N (USA)	13 Oct	Atlas F	3097 lb

Soviet Cosmonauts

44.	Vladimir Dzhanibekov	(Soyuz 27)	10 Jan
27.	Oleg Makarov (3rd flight)		
33.	Alexi Gubarev (2nd flight)	(Soyuz 28)	2 Mar
41.	Vladimir Kovalenok (2nd flight)	(Soyuz 29)	15 Jun
45.	Alexander Ivanchenkov		
28.	Pyotr Klimuk (3rd flight)	(Soyuz 30)	27 Jun
5.	Valery Bykovsky (3rd flight)	(Soyuz 31)	26 Aug

Intercosmos Cosmonauts

1.	Vladimir Remek (Czechoslovakia)	(Soyuz 28)	2 Mar
2.	Miroslaw Hermaszewski (Poland)	(Soyuz 30)	27 Jun
3.	Sigmund Jaehn (East Germany)	(Soyuz 31)	26 Aug

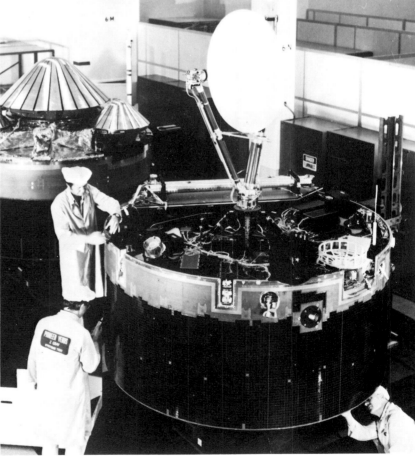

At left: The project manager for the Pioneer Venus mission stands next to a thermal model of the Pioneer Venus 2 Multiprobe (see also *above*).

The large probe with the peaked cap in the middle of the craft and the three smaller probe modules arrayed around it were designed to transmit data only up to the time that they contacted Venus' surface, yet one of these continued to transmit, after landing, for some 67 minutes.

The Pioneer Venus 1 Orbiter (at *right* and at *top* of page) was designed to observe Venus over the duration of one Venusian day (equal to 243 Earth days), and reached Venus one week before the Multiprobe (seen in the *background* here), as the two craft were launched separately.

Above: GOES-C was third in the series of Geostationary Operational Environmental Satellites built by Ford Aeronutronic for NASA. The National Oceanic and Atmospheric Administration NOAA controlled the GOES satellites. GOES-D through -F, which design is seen *at right*, were built by Hughes Aircraft, and were the first spacecraft capable of near-continuous observation of atmospheric water vapor and temperature, for the study of severe storms. *Overleaf:* This weather satellite image of the Central America/Gulf Coast area, taken at 5pm EDT on 26 January 1978, reveals the surprising optical capability of spaceborne systems.

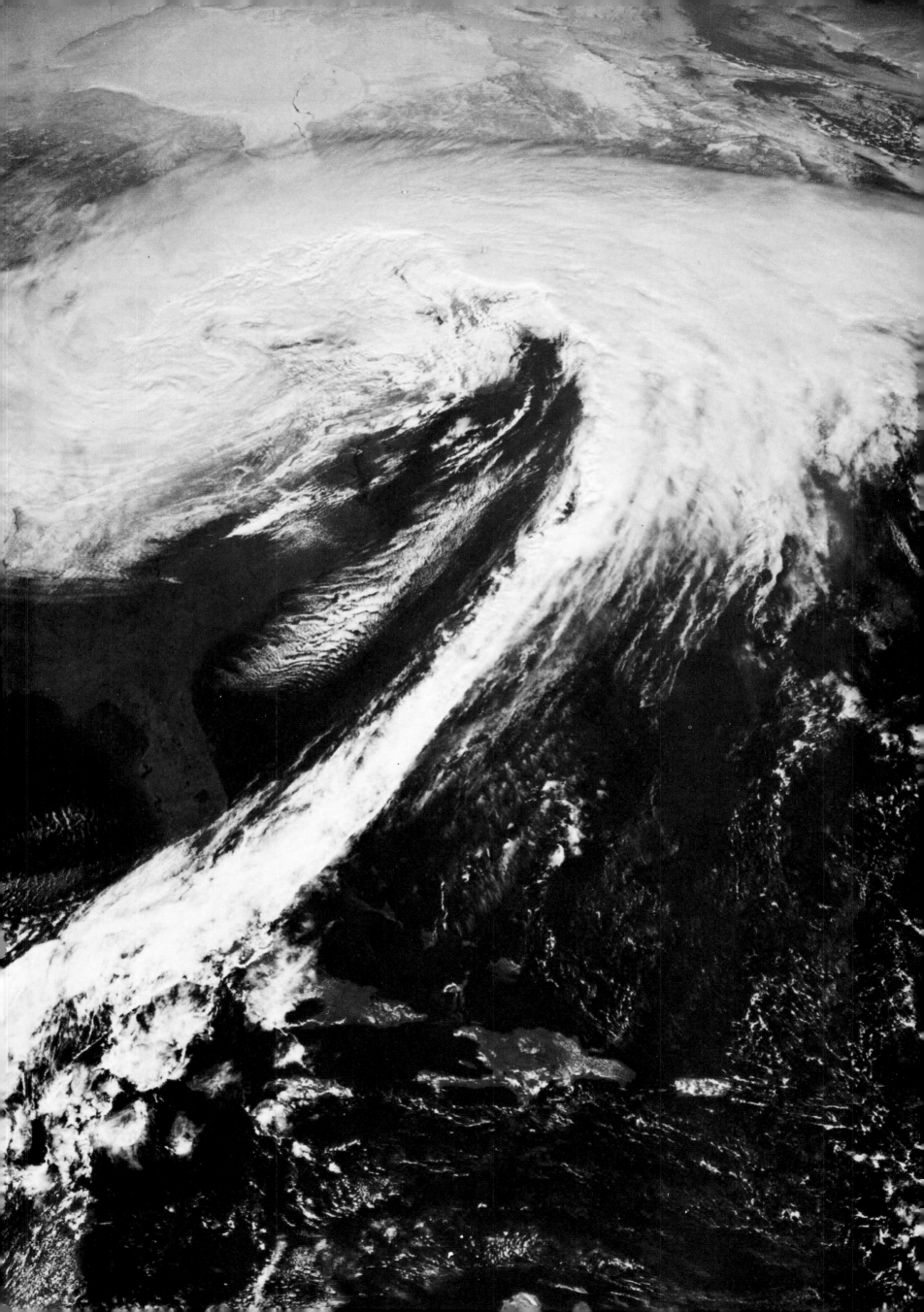

1979

The decade ended with the Soviet Union continuing the spectacular pattern of space station operation that it had begun in 1978, although things didn't go quite as smoothly in 1979. The pattern was renewed on 25 February when Vladimir Lyakhov and Valery Ryumin were launched up to Salyut 6 aboard Soyuz 32. Progress 5 came with supplies and was jettisoned to make way for Soyuz 33, which was launched on 10 April with Soviet cosmonaut Nikolai Rukavishnikov and Bulgarian cosmonaut Georgi Ivanov. For the first time in Salyut 6's heretofore successful career, a Soyuz crew could not dock with the station. The problem lay with the spacecraft and the two cosmonauts were forced to make a successful, if difficult, nighttime emergency landing.

After the near disaster of Soyuz 33 the Soviets decided not to launch any more manned missions to Salyut 6 during the tenure of the Soyuz 32 crew. Faced with the problem that the Soyuz 32 spacecraft might not be functional when its crew was ready to return home, the Soviets made the decision to launch a replacement Soyuz *by remote control*. This launch was successful on 6 June, and on 19 August, Lyakhov and Ryumin returned home aboard Soyuz 34, having set yet another endurance record of 175 days and 36 minutes!

A final visitor to Salyut 6 during 1979 was the unmanned Soyuz T-1, which was launched on 16 December. The prototype of a new, redesigned Soyuz capsule, this spacecraft was the harbinger of the T or 'Troika' series which eventually would be able to carry three cosmonauts in space suits.

A major, though not often remembered, space science effort of 1979 was Cosmos 1129. Launched on 25 September by the Soviet Union with a Cosmos designation, it was actually an international biologic sciences mission with experiments from Bulgaria and the Soviet Union (white rat breeding), Japan (quail eggs) and the United States (carrots and bacteria).

The major American accomplishment in space during 1979, and that which made up for the slippage of the Space Shuttle schedule, came in the payoff from the two Voyager spacecraft that had been launched two years before. Voyager 1 arrived at Jupiter on 5 March and her sister ship followed on 9 July. The fantastic results included breathtaking, detailed, close-up photos of Jupiter's turbulent atmosphere and her awe-inspiring Great Red Spot, a phosphene storm three times larger than the earth that has raged for centuries. The Voyagers also revealed a great many surprises about Jupiter's huge moons, including the fact that Io is the most volcanically active body in the solar system. Voyager's cameras, in fact, photographed eight of Io's sulfurous volcanos *while they were erupting*!

Above: **Soyuz 33 Intercosmonaut Georgi Ivanov, of Bulgaria. Below: Building a Space Shuttle. The Space Shuttle was supposed to make its first powered flight in 1979, but didn't. The Voyager successes took up the slack for NASA, however.** *At right:* **A Soyuz Orbital Module on display in the Soviet Union.**

Space Flight Highlights of 1979

	Launch date	Launch vehicle	Launch weight
Soyuz 32 (USSR)	25 Feb	A-2	14,454 lb
Soyuz 33 (USSR)	10 Apr	A-2	14,454 lb
FLTSATCOM B (USA)	4 May	Atlas/Centaur	4136 lb
Westar 3 (USA)	9 Aug	Delta	674 lb
Cosmos 1129 (USSR)	25 Sep	A-2	12,100 lb

Soviet Cosmonauts
46.	Vladimir Lyakhov	(Soyuz 32)	25 Feb
42.	Valery Ryumin (2nd flight)		
23.	Nikolai Rukavishnikov (2nd flight)	(Soyuz 33)	10 Apr

Intercosmos Cosmonaut
4.	Georgi Ivanov (Bulgaria)	(Soyuz 33)	10 Apr

Opposite: A model of a FLTSATCOM in orbit. FLTSATCOMs relay voice, teletype and computer communications for US land, sea and air military forces; their main duty is as US Navy Fleet Satellite Communications craft for data exchanges between command posts and ships. *Below:* Inspecting the third FLTSATCOM; these military satellites provide more than 30 voice and 12 teletype channels for a wide variety of receivers. *At left:* WESTAR-C is here being encased in its launch shroud.

Jupiter's winds blow for extremely long durations, and yet the winds at various altitudes blow at different speeds, causing the massive planet to have an ever-shifting appearance; for instance, the photo *at above left* was taken by Voyager 1 on 24 January 1979, and the photo *at above right* was taken by Voyager 2 on 9 May 1979. Note the change in position of the famed Great Red Spot, and the differences in atmospheric swirling between the two photos. *Below:* A Voyager 1 image of volcanic, sulfurous Io (at photo left) and icy Europa, as the two moons transitted Jupiter's immense face. Both of these moons are comparable in size to Earth's Moon.

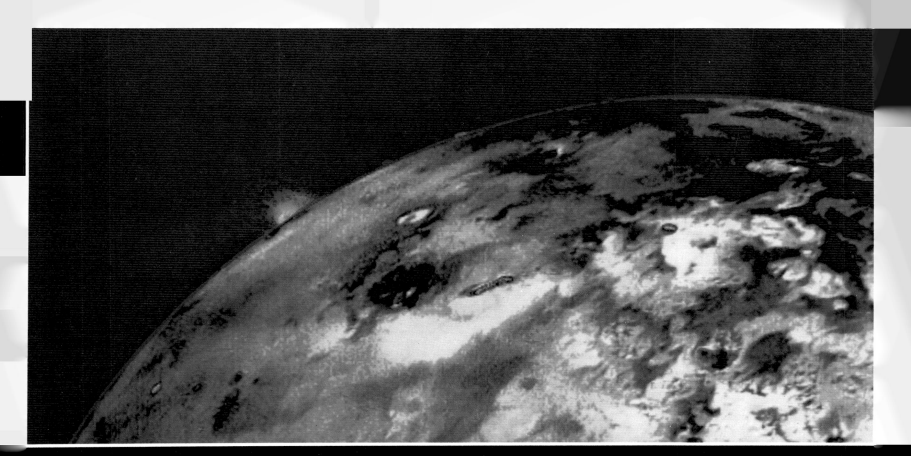

One of the big surprises of the Voyager data return was the extremely active surface of Jupiter's moon Io. *Above* is a Voyager photo of one of the moon's many active volcanoes erupting. Often described as 'pizza-like,' this description applies to appearance only, as Io's red and yellow surface is the product of massive sulfur flows from the moon's molten interior. *Below:* A Voyager photo of Jupiter's equator toward its south pole near the Great Red Spot. Note the large white atmospheric storm just south of the latter, which is itself a massive, centuries-long, storm. Jupiter's liquid, marbled atmosphere contains helium, hydrogen, methane, ammonia, acetylene and other gases.

1980

For the third year running, the United States manned space effort lay dormant while the Soviet Union sent crews to the well-used Salyut 6 space station with almost commuter train-like regularity. The unmanned Progress 8 supply ship was sent up on 27 March, and on 9 April the Soyuz 35 crew was launched. This crew included Leonid Popov and Valery Ryumin, the latter of which had set a 175 day, 36 minute space endurance record in 1979 and had been back on earth for only a little over seven months when he went *back* into space!

These two men were joined by Soviet cosmonaut Valery Kubasov and Hungarian cosmonaut Bertalan Farkas, who were launched aboard Soyuz 36 on 26 May. Farkas was the fourth third-country cosmonaut and the first non-Soviet aboard Salyut 6 since 1978. On 5 June, two days after the Soyuz 36 crew departed in the Soyuz 35 spacecraft, Soyuz T-2 arrived at Salyut 6. The first manned flight of a T-series spacecraft, Soyuz T-2 nevertheless carried just two Soviet cosmonauts on a four-day mission. Next to visit Salyut 6 were Viktor Gorbatko and a Vietnamese cosmonaut, who were launched aboard Soyuz 37 on 23 July for an eight-day mission. They returned to earth in Soyuz 36. On 18 September another international crew, Soviet cosmonaut Yuri Romanenko and Cuban cosmonaut Arnaldo Tamayo Mendez, were launched in the Soyuz 38 spacecraft for an eight-day visit to Salyut 6. On 11 October,

Popov and Ryumin returned to earth, having set *another* space endurance record of 184 days and 20 hours!

Valery Ryumin had now logged *361 days and 22 hours in space* since October 1977!

A final Soviet manned mission for the year, launched on 27 November, was the first flight of a T-series spacecraft with a full three man crew. Visiting Salyut 6 aboard Soyuz T-3 were Leonid Kizim, Oleg Makarov and Gennady Strekalov, the first three man Soyuz crew since the Soyuz 11 disaster nine years before.

For the United States, 1980 was another year of preparing for the first flight of the Space Shuttle Transportation System and watching the spectacular exploits of the Voyager spacecraft in the outer solar system. On 12 November, Voyager 1 passed within 77,000 miles of Saturn, close enough to photograph details of the great planet's rings, now seen to number in the hundreds and to contain graceful, twisting, braided rings within the spaces *between* rings. Having photographed Saturn, Voyager 1 was redirected to fly past her largest moon, Titan. The only moon in the solar system known to possess an atmosphere could now be seen as a dynamic world, where methane rain and snow fall from dense methane clouds, where methane rivers flow through methane channels, and methane oceans fill icy basins.

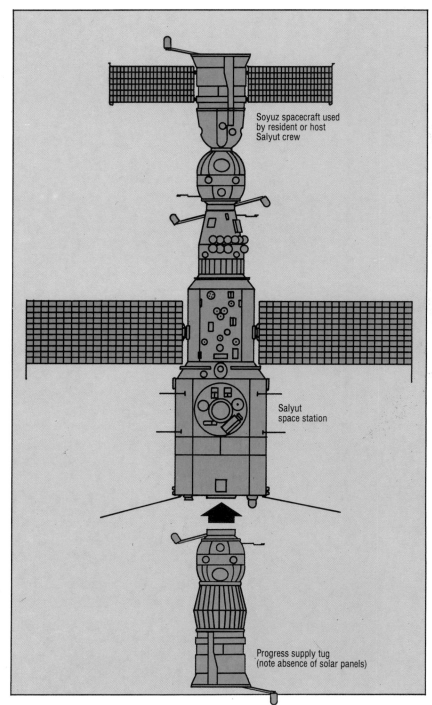

Soyuz spacecraft with a resident or host Salyut crew

Salyut space station

Soyuz spacecraft used by resident or host Salyut crew

Salyut space station

Progress supply tug (note absence of solar panels)

Above: Although this isn't a 1980 photo, the Soviet A-2 launch vehicle shown being transported here was the type of booster used for all Soyuz space missions, including regular 1980 Soyuz missions 35, 36, 37 and 38. The illustrations *across the bottom of* *these pages* show the routine comings and goings of Soyuz spacecraft and Progress supply vehicles at the Salyut 6 space station, which traffic truly attained 'commuter train-like regularity.'

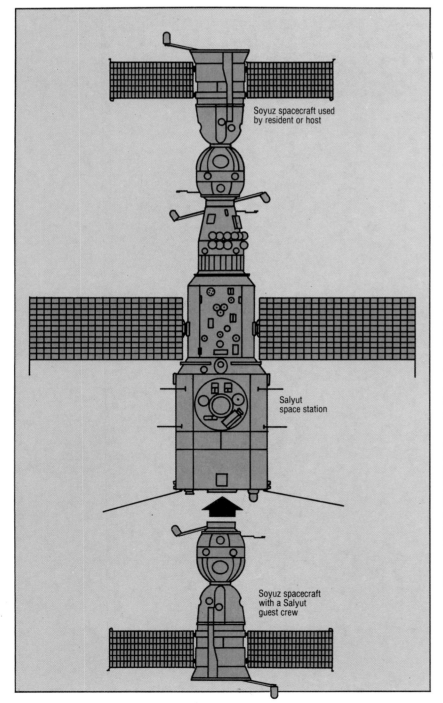

Soyuz spacecraft used by resident or host

Salyut space station

Soyuz spacecraft with a Salyut guest crew

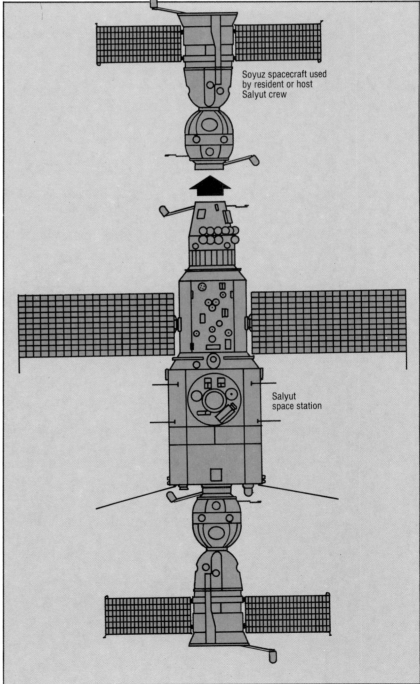

Soyuz spacecraft used by resident or host Salyut crew

Salyut space station

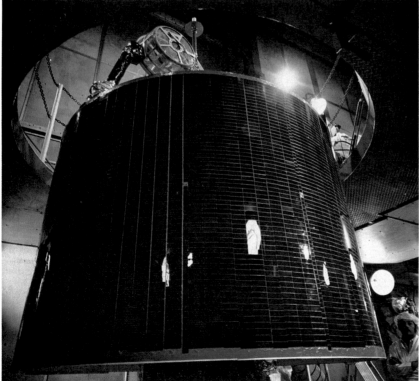

At top of page: Valeri Ryumin and Leonid Popov on board Salyut 6; though the space suit behind Popov seems eerily to be reaching for him, the two cosmonauts were alive and well after a record 184 days and 20 hours in space. *At left:* GOES-D. *Above:* An artist's concept of the Solar Maximum Mission (SMM) craft in space. *At right:* SBS-A, the first of three Satellite Business Systems (SBS) satellites.

Space Flight Highlights of 1980

	Launch date	Launch vehicle	Launch weight
Solar Maximum Mission (USA)	14 Feb	Delta	5093 lb
Soyuz 35 (USSR)	9 Apr	A-2	14,454 lb
Soyuz 36 (USSR)	26 May	A-2	14,454 lb
Soyuz T-2 (USSR)	5 Jun	A-2	15,400 lb
Soyuz 37 (USSR)	23 Jul	A-2	14,454 lb
GOES-D (USA)	9 Sep	Delta	1840 lb
Soyuz 38 (USSR)	18 Sep	A-2	14,454 lb
SBS-1 (USA)	15 Nov	Delta	2325 lb
Soyuz T-3 (USSR)	27 Nov	A-2	15,400 lb

Soviet Cosmonauts

47.	Leonoid Popov	(Soyuz 35)	9 Apr
42.	Valery Ryumin (3rd flight)		
18.	Valery Kubasov (3rd flight)	(Soyuz 36)	26 May
48.	Yuri Malyshev	(Soyuz T-2)	5 Jun
37.	Vladimir Aksyonov (2nd flight)		
21.	Viktor Gorbatko (3rd flight)	(Soyuz 37)	23 Jul
43.	Yuri Romanenko (2nd flight)	(Soyuz 38)	18 Sep
49.	Leonoid Kizim	(Soyuz T-3)	27 Nov
27.	Oleg Makarov (4th flight)		
50.	Gennady Strekalov		

Intercosmos Cosmonauts

5.	Bertalan Farkas (Hungary)	(Soyuz 36)	26 May
6.	Pham Tuan (Vietnam)	(Soyuz 37)	23 Jul
7.	Arnaldo Tamayo Mendez (Cuba)	(Soyuz 38)	18 Sep

At left: SBS-1, ready for its launch shroud. The SBS satellites handle US business communications. *Above:* The Solar Maximum Mission (aka 'Solar Max') satellite is mated to the Multi-Mission Spacecraft (MMS). MMS provided power, control and communications for the Solar Max's solar flare studies. *At right:* Soviet cosmonaut Valery Kubasov *(left)* and Intercosmonaut Bertalan Farkas in survival training.

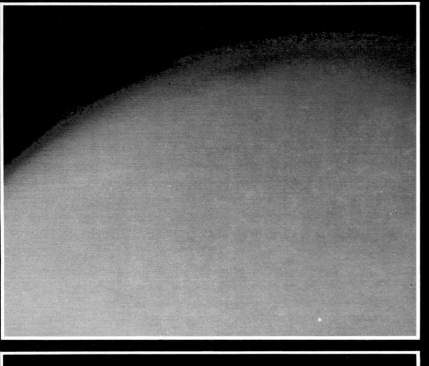

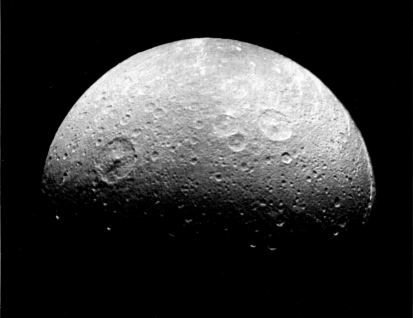

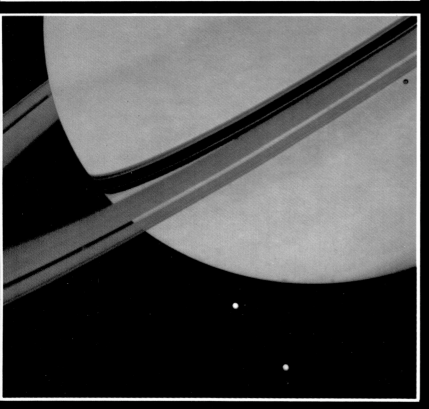

Above, from the top of the page down: Voyager views of Saturn moons: Titan, showing its hazy methane atmosphere; the heavily cratered Dione; Tethys (upper moon) and Dione as they are about to transit Saturn—note the shadows of Tethys and the more prominent rings against the planet's cloud tops. *At right:* A Voyager 1 farewell view of Saturn, four days after flyby. The bright spots which are visible in the rings are the mysterious 'spoke' features.

Part Five

The Age Of The Space Shuttle

1981–1984

1981

After several years of intensive and fruitful manned space flight activity, the Soviet Union passed the baton of manned space flight leadership to the United States, which had launched only *one* manned space mission since 1973, and *none* since 1975.

It began, however, as a Soviet dominated year. Vladimir Kovalenok and Victor Savinykh were launched aboard Soyuz T-4 on 12 March and became the *14th* crew to use Salyut 6 the following day. Savinykh, meanwhile, became the 100th man in space. The two were joined on 23 March by Vladimir Dzhanibekov and Mongolian cosmonaut Jugderdemidyin Gurragcha aboard Soyuz 39 for eight days. On 14 May Soyuz 40 was launched. The last of the original Soyuz series spacecraft, it carried the last international, intercosmos crew—Soviet cosmonaut Leonid Popov and Rumanian cosmonaut Dumitru Prunariu—to go into space, and it is the last spacecraft to dock with Salyut 6. On 26 May, four days after the departure of Soyuz 40, the Soyuz T-4 crew are the last to see the now venerable Salyut 6.

Salyut 6 was at this point the most successful space station in history. It had been in service for nearly four years, hosting 16 manned spacecraft dockings in 18 attempts. It had been manned for a total of 676 days by 28 cosmonauts from eight countries, including six men (Kubasov, Kovalenok, Makarov, Popov, Romanenko and Ryumin) who went aboard twice.

For the Soviet Union, 1981 was the triumph of saying goodbye to a successful and proven space station. For the United States, 1981 was the triumph of inaugurating a new era in spacecraft technology: the Space Shuttle was now ready!

On 12 April, *Columbia* (OV-102), the first of five space-rated Space Shuttle orbiters, was launched into space from the Kennedy Space Center. With astronauts John Young (a veteran of two Gemini and two Apollo missions) and Robert Crippen aboard, the 76-ton *Columbia* completed a 54 hour, 22 minute flight, returning to an airplane-like landing at Edwards AFB, California. The beauty of *Columbia* was that she could fly again!

On 12 November *Columbia* went back into space for a 54 hour, 13 minute flight designated STS-2 for Space Transportation System, second mission.

Another highlight of 1981 for the United States came on 25 August as Voyager 2 made its flyby of the planet Saturn, adding even more to the rich treasury of imagery that its sister ship had gathered.

Below: Space Shuttle *Columbia* rises upon a plume of rocket fire to its second space visit with astronauts Joe Engle and Richard Truly aboard. The white fuel tank was used for 1981 missions only—after that NASA went to an unpainted rust red primer-covered tank to save a thousand pounds of weight. *At right: Columbia*, just prior to its first mission, which was designated 'STS-1.'

Above, left to right: Mission director George Abbey poses as a reading stand so that astronauts Richard Truly and Joe Engle can read their own statements, as reported in a newspaper, concerning the second launch attempt for the weather-delayed *Columbia* STS-2 (Space Shuttle Transportation System, second) mission. *These pages:* Robert Crippen addresses the press after the *Columbia* STS-1 mission return. Mission commander John Young is standing with his wife, Susy, to the right of STS-1 pilot Crippen, and Robert Crippen's own wife, Virginia, and family are standing behind the Youngs. America's space plane truly became a space plane with STS-1.

Space Flight Highlights of 1981

	Launch date	Launch vehicle	Launch weight
Soyuz T-4 (USSR)	12 Mar	A-2	15,400 lb
Soyuz 39 (USSR)	22 Mar	A-2	14,454 lb
Soyuz 40 (USSR)	14 May	A-2	14,454 lb

American Space Shuttle Missions
Columbia (OV-102)

1.	12 Apr	(STS-1)	Robert Crippen John Young
2.	12 Nov	(STS-2)	Joe Engle Richard Truly

Soviet Cosmonauts

41.	Vladimir Kovalenok (3rd flight)	(Soyuz T-4)	12 Mar
51.	Victor Savinykh		
44.	Vladimir Dzhanibekov (2nd flight)	(Soyuz 39)	22 Mar
47.	Leonid Popov (2nd flight)	(Soyuz 40)	14 May

Intercosmos Cosmonauts

8.	Jugderdemidyin Gurragcha (Mongolia)	(Soyuz 39)	22 Mar
9.	Dumitru Prunariu (Romania)	(Soyuz 40)	14 May

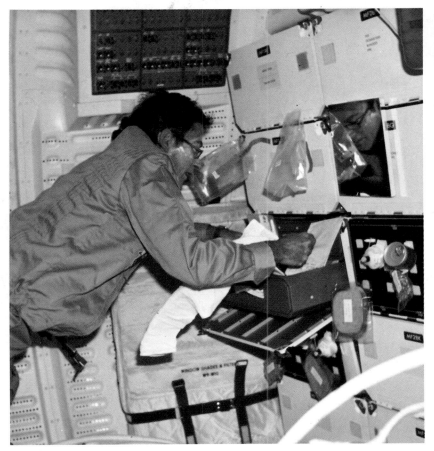

At left: The *Columbia* Orbital Vehicle is being positioned for connection to its huge fuel tank and booster system. *Above:* Astronaut John Young dries his razor after shaving aboard the orbital STS-1. *At right:* STS-1 blasts off en route to becoming the first US manned space flight to return upon solid ground. *Below:* Richard Truly strolls to an STS-2 test.

Below: The Soyuz Orbital Module is part of the three-module Soyuz manned spacecraft system. In flight configuration, attached to this Orbital Module would be the bell-shaped Descent Module, and to that would be attached the Instrument Module—which also carries the Soyuz craft's solar panels. Cosmonauts enter the Descent Module for return to Earth, and eject the Orbital and Instrument modules just after reentry pattern retro-rocket burn. The large, tub-like apparatus at the top of the OM here is an early Soyuz docking adaptor. Later Soyuz dockings were effected with a greatly streamlined, and much reduced in size, docking apparatus. The docking apparatus for the Apollo-Soyuz Test Project (see '1975') modified both Soyuz *and* Apollo.

Above: A conception of Voyager 2 at its closest approach to Saturn and that planet's vast ring system. Below: The moon Enceladus from Voyager 2. At right: A Voyager 2 image of Saturn and moons (from left) Tethys (with dark shadow), Dione and Rhea; Mimas is a light dot on Saturn's face to Tethys' upper left, and its shadow is close under the rings—which here can be seen to evidence 'spokes.'

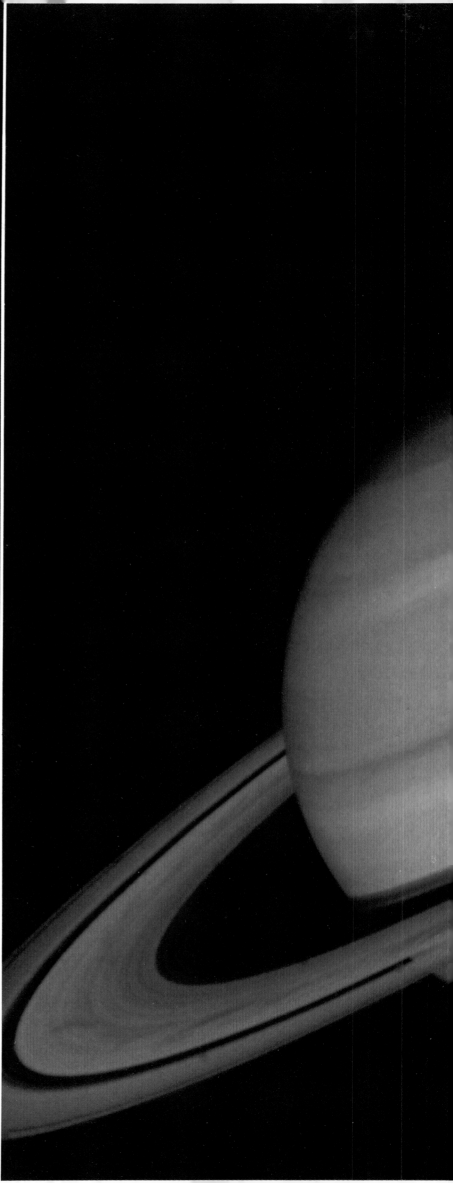

1982

Having made its two initial flights in 1981, the Space Shuttle Orbiter *Columbia* (OV-102) continued its four-flight test program with the eight-day STS-3 mission, that was launched on 22 March, and the seven day STS-4 on 27 June. On 11 November, *Columbia* was launched an incredible *fifth* time. This mission, STS-5, carried a four man crew and marked the first 'operational' STS flight as *Columbia* launched two satellites from its payload bay—the American SBS-3 and the Canadian Anik-C.

Meanwhile, on 19 April, the Soviet Union launched Salyut 7, the successor to the historic Salyut 6. A month later, on 14 May, the two man crew of Soyuz T-5 arrived to inaugurate manned occupancy of the new space station.

On 25 June, the crew of Soyuz T-6 arrived at the new space station. Aboard was a two man Soviet crew and French cosmonaut Jean-Loup Chretien, the first non-Soviet cosmonaut who was also not a member of the Intercosmos program. Another eight-day mission to Salyut 7 came on 19 August carrying three Soviet cosmonauts, among them Svetlana Savitskaya, the second woman in space and the first since Valentina Tereshkova's historic 1963 flight. In keeping with the pattern established in Soyuz flights of earlier years, the crew departed aboard Soyuz T-5, leaving the newer Soyuz T-7 at the space station.

The DSCS (aka 'Discus') satellites are designed to provide the US Department of Defense with a reliable, global-coverage, strategic communications network. DSCS-3 *(below)* was built with better anti-jam protection than DSCS-1 and DSCS-2, and had, in addition, 50 percent greater communications capability than the latter. DSCS satellites are managed by the US Air Force.

Below opposite: This Anik-C communications satellite is one of three built by Hughes Aircraft for Telesat of Canada. Meant to keep pace with Canada's growing communications needs over the next decade, the first of these was launched from the Space Shuttle *Columbia's* payload bay during Shuttle mission STS-5. 'Anik' is said to be Eskimo for 'brother.'

SATELLITE DEPLOYMENT BY

Ace Moving Co.

FAST AND COURTEOUS SERVICE

"WE DELIVER"

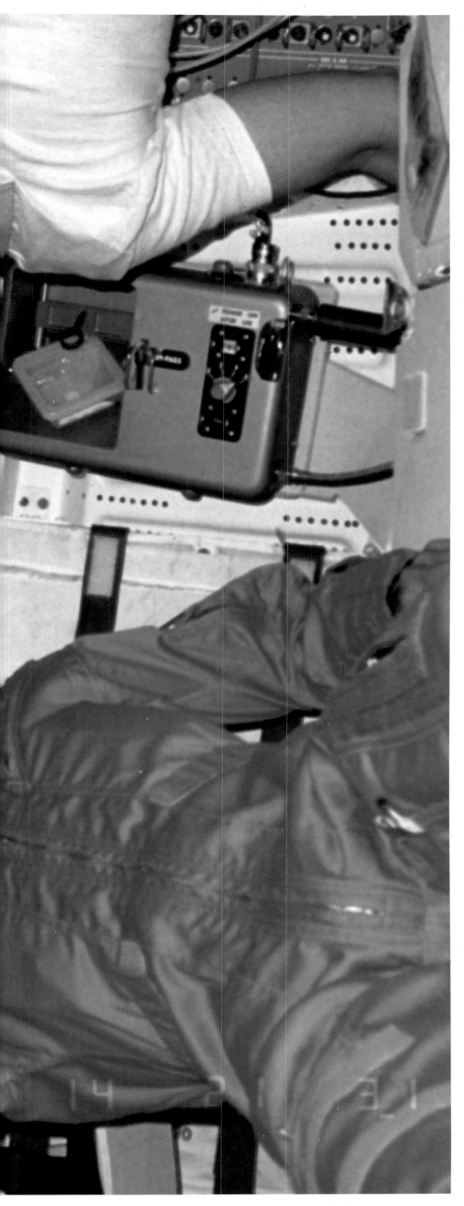

At left: The crew of STS-5, the first Shuttle flight to launch satellites from its capacious payload bay—clockwise, starting with sign-holder Vance D Brand, William B Lenoir, Robert F Overmyer and Joseph P Allen IV. Satellites deployed were an Anik-C (see pages 150-151) and the American SBS-3, a business communications satellite which was built by Hughes for small business systems. SBS-3's services proved to be in high demand. *Above:* The Shuttle *Columbia* blasts off.

Space Flight Highlights of 1982

	Launch date	Launch vehicle	Launch weight
Soyuz T-5 (USSR)	13 May	A-2	15,400 lb
Soyuz T-6 (USSR)	24 Jun	A-2	15,400 lb
Soyuz T-7 (USSR)	19 Aug	A-2	15,400 lb
DSCS-3 (USA)	30 Oct	Titan	1876 lb
Anik-C (Canada)	11 Nov	Shuttle	1250 lb

American Space Shuttle Missions
Columbia (OV-102)

1.	22 Mar	(STS-3)	Jack Lousma
			Gordon Fullerton
2.	27 Jun	(STS-4)	Henry Hartsfield, Jr.
			Thomas Mattingly II
3.	11 Nov	(STS-5)	Joseph Allen
			Vance Brand
			William Lenoir
			Robert Overmyer

Soviet Cosmonauts

52.	Anatoli Berezovoi	(Soyuz T-5)	13 May
29.	Valentin Lebedev		
	(2nd flight)		
44.	Vladimir Dzhanibekov	(Soyuz T-6)	24 Jun
	(3rd flight)		
45.	Alexander Ivanchenkov		
	(2nd flight)		
47.	Leonoid Popov	(Soyuz T-7)	19 Aug
	(3rd flight)		
53.	Alexander Serebrov		
54.	Svetlana Savitskaya		

French Cosmonaut

1.	Jean-Loup Chretien	(Soyuz T-6)	24 Jun

At left: STS-5 astronaut Joseph P Allen's spot meter floats handily near, during an image-making session devoted to planet Earth (as photographed from *Columbia's* windows). *Above:* Jean-Loup Chretien gained distinction as France's first man in space, aboard Soyuz T-6/Salyut 7. *Below:* Salyut 7.

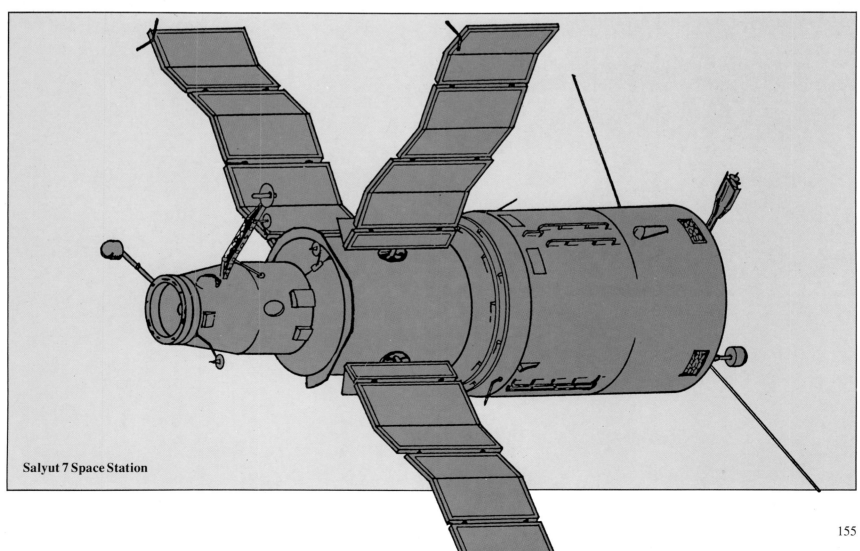

Salyut 7 Space Station

155

Above: The round capsule atop this Venera probe houses the craft's Descent Module. In 1982, Venera 13/14 sent the first color pictures of Venus' hot, pressure-intensive surface *(below and at bottom)*; these were fisheye lens shots—note *at above right* the lander's shock absorbing base, an ejected lens cap and the Venusian horizon. *Below right:* Another view.

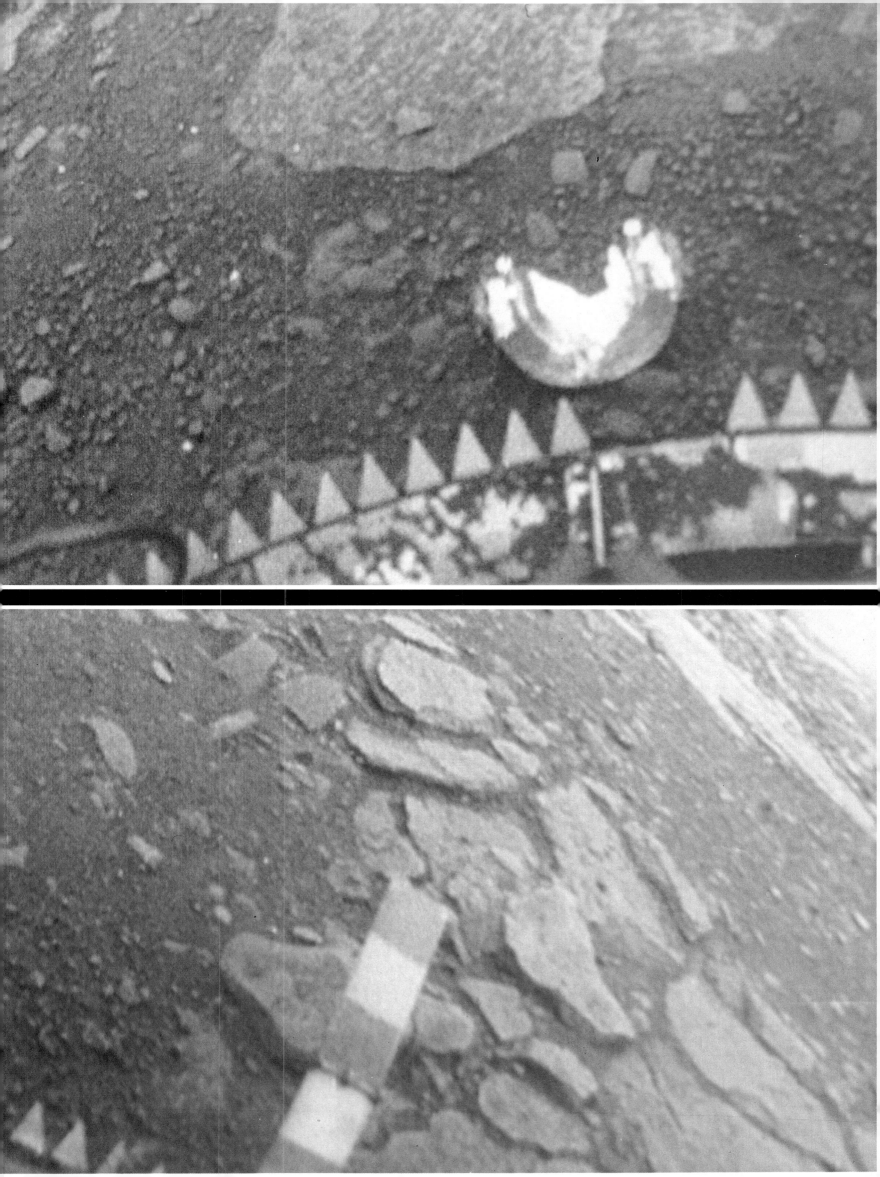

1983

For the United States 1983 was marked by a slow, yet steady, expansion of the Space Shuttle program, with four flights scheduled during the year. NASA also brought on line the second of a planned fleet of four Space Shuttle Orbiters—the 75-ton *Challenger* (OV-99).

It was *Challenger*, in fact, that flew the first three STS missions of the year: STS-6 launched for 5 days on 4 April; STS-7 launched for 6 days on 18 June; and STS-8 launched for 6 days on 30 August. During *Challenger's* debut season Sally Ride became the first American woman in space (STS-7) and Guion Bluford became the first black American in space (STS-8).

STS-7 also marked the first use of the free-flying West German SPAS module that was used to take the 70mm color photos seen on these pages of the Shuttle Orbiter in space. Mission STS-8 in turn was the first nighttime launch of the Space Shuttle Transportation System.

The fourth STS mission of 1983 was the sixth flight for *Columbia*, and the first operational use of the European Space Agency (ESA) Spacelab Module, which was designed to fit within the entire length of the Orbiter's payload bay. Launched on 28 November, STS-9 was, at 10 days, the longest mission that a Space Shuttle Orbiter would fly before the spacecraft system was withdrawn from service in 1986. Among the six man crew—the largest ever carried by a spacecraft—was the first non-American to fly aboard an American spacecraft. Officially designated as a payload specialist rather than an astronaut, West Germany's Ulf Merbold was the official ESA representative for the first flight of the Spacelab scientific experiment module.

The Soviet year in manned space flight centered once again around the Salyut 7 space station. Soyuz T-8, the first attempt to man the station during 1983, was launched on 20 April but the crew failed in their docking maneuver and were forced to return to earth after two days. The second attempt, Soyuz T-9, followed on 27 June. Vladimir Lyakhov and Alexander Alexandrov managed a successful docking and moved into their new home. On 27 September, Vladimir Titov and Gennady Strekalov, the crew of the abortive Soyuz T-8 mission, tried again aboard Soyuz T-10. This time the launch vehicle exploded on the launch pad. Fortunately the cosmonauts had fired the spacecraft's rocket-powered escape system and they survived.

Meanwhile aboard Salyut 7, Alexandrov and Lyakhov had some troubles of their own. A leak of toxic oxidizer from the space station's propellant tanks forced them to briefly consider emergency evacuation on 9 September, but the situation stabilized and they remained in space until 23 November.

Below: **Dr Sally K Ride became the first American woman in space aboard the *Challenger* STS-7 mission. Note the calculators, papers and food container floating around Dr Ride, who also floats in zero G. Also, note the view of Earth out the Shuttle Orbiter cabin windows behind Dr Ride's head.** *At right: Challenger* **STS-7.**

Opposite: The West German pallet satellite SPAS-1, which has here just become the first satellite launched with the Shuttle's remote manipulator arm, took this view of *Challenger* STS-7. *At left:* STS-9 astronaut John W Young aboard *Columbia. Above:* STS-9's West German payload specialist, Ulf Merbold. *Below:* Preparing for the historic night launching of a Shuttle Orbiter is *Challenger* STS-8, which also carried the first black American astronaut, Guion Bluford.

At right: STS-8 was the first Space Shuttle mission to carry the Payload Flight Test Article. *At top, above:* The Canadian Anik-C2 satellite, as seen from, and freshly launched from, STS-7. *Above:* STS-8's Payload Flight Test Article. *At right:* Indonesia's Palapa-B satellite, (a Hughes HS 376), upon its STS-7 launching.

Telstar 3 Spacecraft

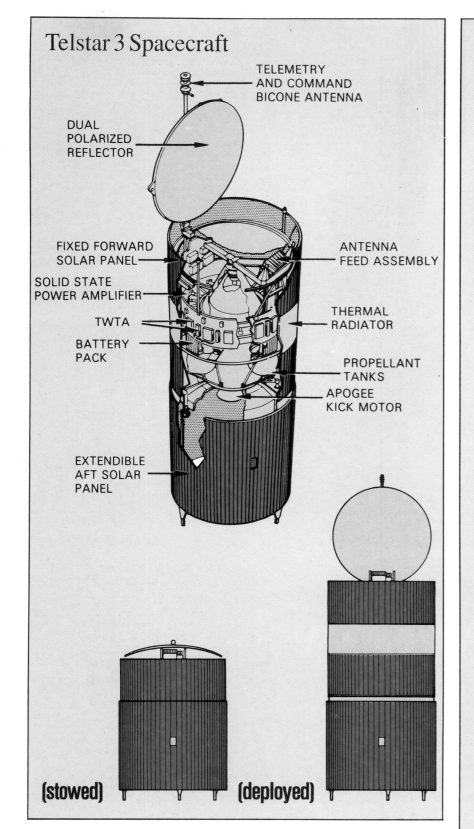

- TELEMETRY AND COMMAND BICONE ANTENNA
- DUAL POLARIZED REFLECTOR
- FIXED FORWARD SOLAR PANEL
- SOLID STATE POWER AMPLIFIER
- TWTA
- BATTERY PACK
- EXTENDIBLE AFT SOLAR PANEL
- ANTENNA FEED ASSEMBLY
- THERMAL RADIATOR
- PROPELLANT TANKS
- APOGEE KICK MOTOR

(stowed)　　　(deployed)

At left: A Hughes Aircraft Company technician adjusts a travelling wave tube amplifier on Telstar 3 (see also *above*), prior to the satellite's launch from Kennedy Space Center on 28 July 1983. The Telstar designation was revived by AT&T and was given to a series of Hughes HS 376 satellites which were bought by AT&T. *Below:* Having just completed its first mission (STS-6), the Shuttle Orbiter *Challenger* touches down at Edwards Air Force Base.

Space Flight Highlights of 1983

	Launch date	Launch vehicle	Launch weight
IRAS (International)	25 Jan	Delta	2249 lb
Soyuz T-8 (USSR)	20 Apr	A-2	15,400 lb
Anik-C2 (Canada)	18 Jun	Shuttle	1250 lb
Palapa-B (Indonesia)	19 Jun	Shuttle	1437 lb
Soyuz T-9 (USSR)	27 Jun	A-2	15,400 lb
Telstar 3A (USA)	28 Jul	Delta	7000 lb
Galaxy 3 (USA)	21 Sep	Shuttle	1222 lb

American Space Shuttle Missions
Challenger (OV-99)

1.	4 Apr	(STS-6)	Karol Bobko
			Story Musgrave
			Donald Peterson
			Paul Weitz
2.	18 Jun	(STS-7)	Robert Crippen
			John Fabian
			Frederick Hauck
			Sally Ride
			Norman Thagard
3.	30 Aug	(STS-8)	Guion Bluford, Jr
			Daniel Brandenstein
			Dale Gardner
			William Thornton
			Richard Truly

Columbia (OV-102)

1.	28 Nov	(STS-9 (41-A)/ Spacelab 1)	Owen Garriott
			Byron Lichtenberg
			Robert Parker
			Brewster Shaw, Jr
			John Young

Soviet Cosmonauts

55.	Vladimir Titov	(Soyuz T-8)	20 Apr
50.	Gennady Strekalov (2nd flight)		
53.	Alexander Serebrov (2nd flight)		
46.	Vladimir Lyakhov (2nd flight)	(Soyuz T-9)	27 Jun
56.	Alexander Alexandrov		

ESA Payload Specialist

1.	Ulf Merbold (West Germany)	(STS-9 (41-A)/ Spacelab 1)	28 Nov

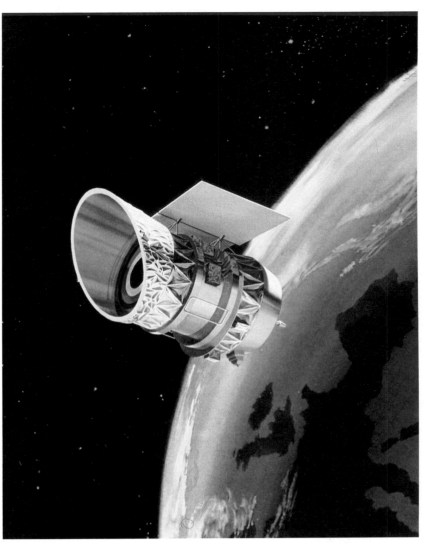

The Infrared Astronomical Satellite (IRAS) was a joint project of NASA and the Netherlands, with British participation. IRAS conducted deep space studies. *These pages,* clockwise from lower left: IRAS in construction at the Fokker facility in the Netherlands; IRAS in testing; IRAS being fitted with its launch shroud; technicians ready IRAS for launching; an artist's concept of IRAS in orbit; and the IRAS launch on 25 January 1983.

These pages: A SPAS-1 view of Shuttle Orbiter *Challenger* during STS-7. The *Challenger's* Remote Manipulator Arm is visible on the right side of the payload bay. Before returning to Earth, *Challenger's* crew retrieved the SPAS-1 satellite with the manipulator arm, and stowed the SPAS-1 in the payload bay—from whence it came. Visible also are the vacated cradles (which open and shut transversely—see pages 172–174) for the *Challenger*-launched (Canadian) Anik-C2 and (Indonesian) Palapa-B communications satellites, and the pallet for the second NASA Office of Space and Terrestrial Applications (OSTA-2) earth resources instrument bed.

1984

Once again the United States carefully expanded its Space Shuttle Transportation System, with five flights in 1984, compared with four in 1983, three in in 1982 and two in 1981. As in 1983, a new orbiter—the 73.9 ton *Discovery* (OV-103)— was unveiled for its first two operational flights. Fully expecting that the shuttle flights would soon become a routine, almost bimonthly, affair, NASA adopted a cumbersome new designation system for the flights. The three digit designation was composed of a numeral indicating the fiscal year during which the mission was originally scheduled and a digit identifying the launch site. (Kennedy Space Center in Florida was 1 and Vandenberg AFB in California—from which no flight would ever be made while this designation system was in use—would have been 2.) The third digit indicated the place of a particular flight in the sequence of a particular fiscal year. Mission STS-9 was redesignated as Mission 41-A as it was the first flight of fiscal year 1984. (NASA's fiscal year begins in November.)

The first flight of calendar year 1984 was Mission 41-B on which *Challenger* was launched on 3 February. During the eight-day mission, the five man crew launched three satellites, although two of them failed to orbit properly. *Challenger* also flew Mission 41-C, which was launched on 6 April. During this seven day mission the five man crew located and recovered the malfunctioning Solar Maximum Mission satellite. Astronauts George Nelson and James van Hoften then went into *Challenger's* payload bay and successfully repaired the big solar observation spacecraft!

The next American shuttle mission, 41-D, was the maiden voyage of the third space-rated orbiter *Discovery* (OV-103). Launched on 30 August for a five-day flight, *Discovery's* crew of six included Judy Resnik, America's second woman in space. Highlights of Mission 41-D included the launch of three satellites: Leasat 2, Telstar 3 and SBS-4. With Missions 41-E and 41-F canceled, Mission 41-G followed on 5 October. Aboard *Challenger* for this flight were Marc Garneau, a Canadian, Kathryn Sullivan, the third American woman in space, and Sally Ride, who became the first American woman to fly in space twice.

Discovery's nine-day mission, launched on 8 November, was its second and the last American mission for the year. Designated 51-A, it was actually the first of FY 1985. The Mission 51-A crew launched two satellites of its own as well as 'rescuing' the two satellites launched by 41-B that had not orbited properly.

On 8 February, the Soviet Union launched its Soyuz T-10B (replacing the original Soyuz T-10 that was lost on the launch pad in September 1983). Aboard the ship were the first three man crew to reach Salyut 7 in over a year. On 3 April they received their first guest crew, who arrived aboard Soyuz T-11. Among the Soyuz T-11 crew was Rakesh Sharma, India's first cosmonaut. During their own tenure, the Soyuz T-10B crew entered the record books for having logged 35 hours of spacewalk time outside the space station. This compared to 28 hours total spacewalk time for the *entire* Soviet manned space flight program through the end of 1983.

On 18 July, Soyuz T-12 arrived at Salyut 7 with another guest crew that included Svetlana Savitskaya, who became the first Soviet woman to fly in space twice.

On 2 October the original Soyuz T-10B crew of Oleg Atkov, Leonid Kizim and Vladimir Solovyev returned to earth with not only a record for spacewalk hours but a new space endurance record—*of 237 days*—as well. The Soviet Union had logged a total of 3650 cosmonaut days in space, versus the American aggregate of 1250.5 that had been logged through the end of STS Mission 41-D.

Below: **Shuttle Orbiter** *Discovery's* **maiden launch, on mission 41-D—under the new NASA designation (see text).** *At right:* **The** *Discovery* **41-D crew, clockwise from top center: Judith A Resnik, Steven A Hawley, Michael L.Coats, Henry W Hartsfield Jr, Richard M Mullane and Charles D Walker.**

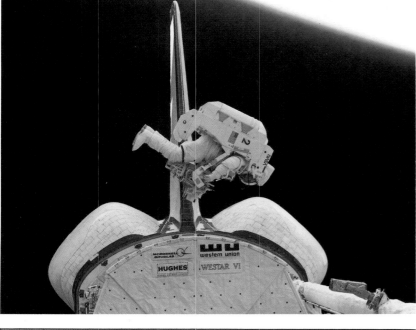

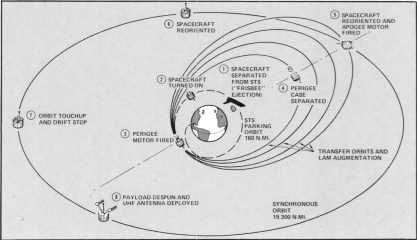

Leasat Spacecraft

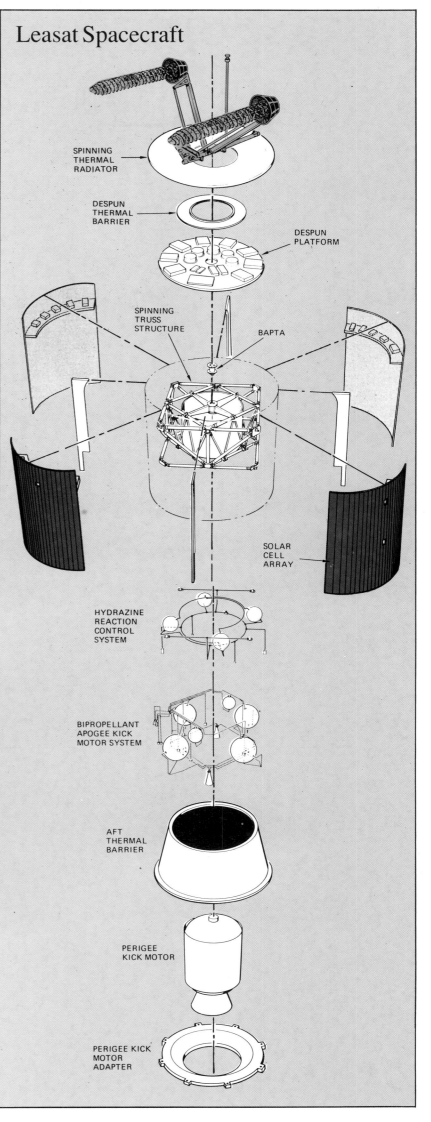

SPINNING THERMAL RADIATOR

DESPUN THERMAL BARRIER

DESPUN PLATFORM

SPINNING TRUSS STRUCTURE

BAPTA

SOLAR CELL ARRAY

HYDRAZINE REACTION CONTROL SYSTEM

BIPROPELLANT APOGEE KICK MOTOR SYSTEM

AFT THERMAL BARRIER

PERIGEE KICK MOTOR

PERIGEE KICK MOTOR ADAPTER

Opposite page—above—The Syncom 4/Leasat 2 satellite just after launch from *Discovery*/41-D *and—below*—leaving its cradle in the payload bay. *At top left:* A Leasat in testing. *At left:* The Leasat launch sequence. *Above:* Leasat. *Above left:* Logos of Westar 4's contributees decorate its cradle: 41-B astronaut Bruce McCandless tests satellite repair equipment which includes the Manned Maneuvering Unit (MMU).

At left: Its cradle opened, SBS-4 is launched from the Space Shuttle in August of 1984. *Above:* SBS-4 heads for orbit. *At right:* The SBS-4 in diagram. *Below:* The method of launching satellites from the Space Shuttle.

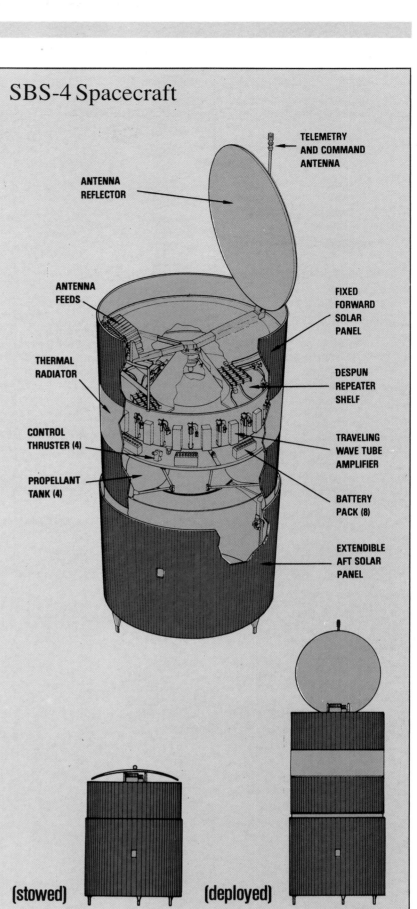

SBS-4 Spacecraft

TELEMETRY AND COMMAND ANTENNA

ANTENNA REFLECTOR

ANTENNA FEEDS

THERMAL RADIATOR

CONTROL THRUSTER (4)

PROPELLANT TANK (4)

FIXED FORWARD SOLAR PANEL

DESPUN REPEATER SHELF

TRAVELING WAVE TUBE AMPLIFIER

BATTERY PACK (8)

EXTENDIBLE AFT SOLAR PANEL

(stowed)   (deployed)

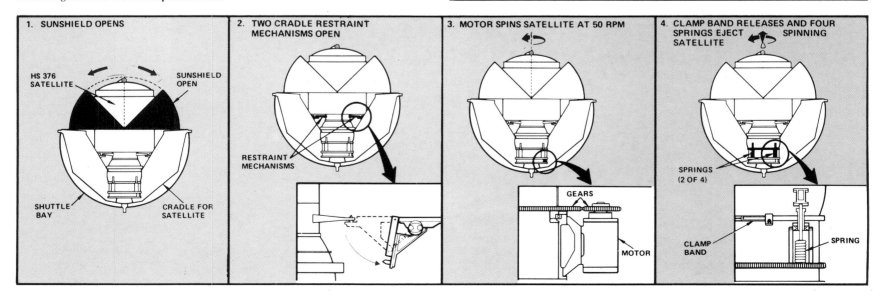

1. SUNSHIELD OPENS

HS 376 SATELLITE

SUNSHIELD OPEN

SHUTTLE BAY

CRADLE FOR SATELLITE

2. TWO CRADLE RESTRAINT MECHANISMS OPEN

RESTRAINT MECHANISMS

3. MOTOR SPINS SATELLITE AT 50 RPM

GEARS

MOTOR

4. CLAMP BAND RELEASES AND FOUR SPRINGS EJECT SPINNING SATELLITE

SPRINGS (2 OF 4)

CLAMP BAND

SPRING

Below: Now, for our Canadian readers, the Canadian Astronaut Program astronauts, clockwise from left rear: Bjarni Tryggvason, Robert Thirsk, Roberta Bondar, Steve MacLean, Marc Garneau (who flew with STS 41-G) and Ken Money. *Above left:* Bruce McCandless enacting the first operational use of the Manned Maneuvering Unit (during mission 41-B), and *above*, riding *Challenger's* Remote Manipulator Arm (RMA). *At right:* A Japanese launch of their GMS-3 satellite.

Space Flight Highlights of 1984

	Launch date	Launch vehicle	Launch weight
Soyuz T-10B (USSR)	8 Feb	A-2	15,400 lb
Soyuz T-11 (USSR)	3 Apr	A-2	15,400 lb
Soyuz T-12 (USSR)	17 Jul	A-2	15,400 lb
GMS-3 (Japan)	3 Aug	N-II	770 lb
Leasat-2 (USA)	1 Sep	Shuttle	2900 lb
SBS-4 (USA)	30 Aug	Shuttle	7000 lb
Vega 1 (USSR)	15 Dec	Proton SL-12	
Vega 2 (USSR)	21 Dec	Proton SL-12	

American Space Shuttle Missions
Challenger (OV-99):

1.	3 Feb	(41-B)	Vance Brand
			Robert Gibson
			Bruce McCandless
			Ronald McNair
			Robert Stewart
2.	6 Apr	(41-C)	Robert Crippen
			Terry Hart
			George Nelson
			Francis Scobee
			James van Hoften
3.	5 Oct	(41-G)	Robert Crippen
			David Leetsma
			Jon McBride
			Paul Scully-Power
			Sally Ride
			Kathryn Sullivan

Discovery (OV-103):

1.	30 Aug	(41-D)	Michael Coats
			Henry Hartsfield, Jr
			Steven Hawley
			Richard Mullane
			Judy Resnik
			Charles Walker
2.	8 Nov	(51-A)	Joseph Allen
			Anna Fisher
			Dale Gardner
			Frederick Hauck
			David Walker

Soviet Cosmonauts

49.	Leonoid Kizim (2nd flight)	(Soyuz T-10B)	8 Feb
57.	Vladimir Solovyev		
58.	Oleg Atkov		
48.	Yuri Malyshev (2nd flight)	(Soyuz T-11)	3 Apr
50.	Gennady Strekalov (3rd flight)		
44.	Vladimir Dzhanibekov	(Soyuz T-12)	17 Jul
59.	Igor Volk		
54.	Svetlana Savitskaya (2nd flight)		

Other Astronauts and Cosmonauts

1.	Marc Garneau (Canada)	(41-G)	5 Oct
2.	Rakesh Sharma (India)	(Soyuz T-11)	3 Apr

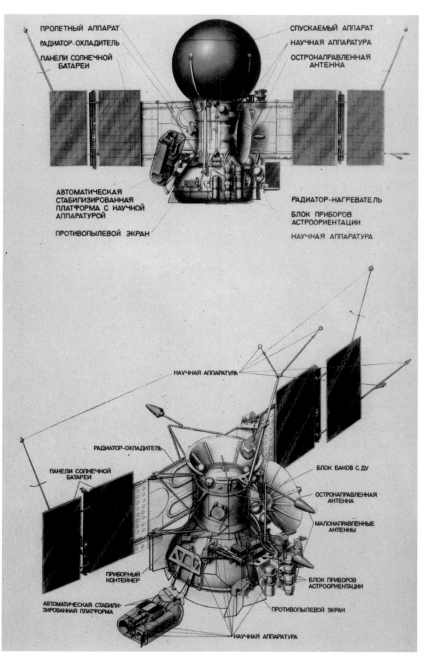

ПРОЛЕТНЫЙ АППАРАТ
РАДИАТОР-ОХЛАДИТЕЛЬ
ПАНЕЛИ СОЛНЕЧНОЙ БАТАРЕИ
СПУСКАЕМЫЙ АППАРАТ
НАУЧНАЯ АППАРАТУРА
ОСТРОНАПРАВЛЕННАЯ АНТЕННА
АВТОМАТИЧЕСКАЯ СТАБИЛИЗИРОВАННАЯ ПЛАТФОРМА С НАУЧНОЙ АППАРАТУРОЙ
ПРОТИВОПЫЛЕВОЙ ЭКРАН
РАДИАТОР-НАГРЕВАТЕЛЬ
БЛОК ПРИБОРОВ АСТРООРИЕНТАЦИИ
НАУЧНАЯ АППАРАТУРА

НАУЧНАЯ АППАРАТУРА
РАДИАТОР-ОХЛАДИТЕЛЬ
ПАНЕЛИ СОЛНЕЧНОЙ БАТАРЕИ
ПРИБОРНЫЙ КОНТЕЙНЕР
АВТОМАТИЧЕСКАЯ СТАБИЛИ-ЗИРОВАННАЯ ПЛАТФОРМА
НАУЧНАЯ АППАРАТУРА
БЛОК БАКОВ С ДУ
ОСТРОНАПРАВЛЕННАЯ АНТЕННА
МАЛОНАПРАВЛЕННЫЕ АНТЕННЫ
БЛОК ПРИБОРОВ АСТРООРИЕНТАЦИИ
ПРОТИВОПЫЛЕВОЙ ЭКРАН

At left: Discovery 51-A Astronauts Dale Gardner (with sign) and Joseph P Allen make light of the Westar 6 (below them) rescue and stowing. *Below:* Allen, on the Remote Manipulator Arm, and Gardner berth the rescued Palapa-B2 satellite. *Above:* Soviet Vegas 1 and 2 were Venus/Halley's Comet probes.

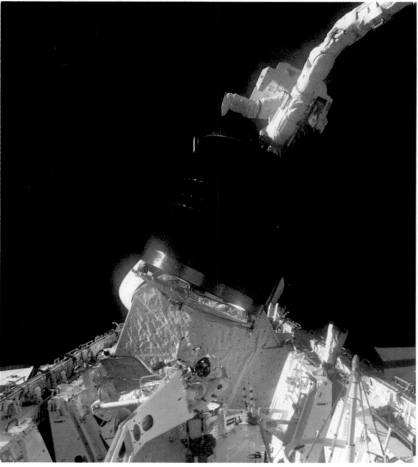

Part Six

Space Flight Comes Of Age

1985–1987

1985

I t was a year of a record number of space flights for a single nation in one year as the United States Space Shuttle Transportation System astronauts commuted into space nine times during 1985. For the third year in a row, NASA also introduced a new Space Shuttle Orbiter: the 74-ton *Atlantis* (OV-104).

The first STS mission of 1985 saw *Discovery* launched on 24 January, carrying a classified Defense Department reconnaissance satellite. *Discovery* also flew the next shuttle mission, which was launched on 12 April. The year's third STS mission, featuring *Challenger* this time, went up on 29 April, marking the first time that *two* shuttle missions had been launched in the same month. This mission, designated 51-B, carried the ESA Spacelab module on its second mission in space.

On 17 June, *Discovery* began its third flight of the year. Designated 51-G, this was the first space flight in history to carry personnel from three countries. In addition to five Americans, the crew included France's Patrick Baudry and Prince Salman Abdel-Aziz Al-Saud of Saudia Arabia. On 29 July, about a month after *Discovery* returned, *Challenger* was launched with the ESA Spacelab. It was Spacelab's third mission and the second of three to be flown aboard *Challenger*.

Discovery began another seven-day flight on 27 August and the new orbiter *Atlantis* made her debut in a four day mission launched on 3 October. The next STS mission belonged to *Challenger* and was launched on 30 October. For the first time three different orbiting vehicles had been in space in less than two months. This mission, 61-A, was the first of FY 1986 and carried the ESA Spacelab module for the third time. Spacelab was designated D-1 for this mission because it was dedicated to West German scientific projects. Among the crew were one Dutch and two German payload specialists.

Atlantis flew the ninth and final STS mission of 1985, which was launched on 27 November. *Columbia* was intended to fly in December, but the flight was postponed to January 1986. Had *Columbia* made its launch date, 1985 would have seen ten STS flights, double the number of 1984 missions, and it would have been the only year in which *all four* of the original space-rated orbiters had made operational flights.

While the United States was having its most prolific year in space ever, the Soviet Salyut 7 space station lay dormant for nearly half a year until Soyuz T-13 arrived on 6 June. On 18 September they were joined by the three man crew of Soyuz T-14. On 26 September, Soyuz T-13 commander Vladimir Dzhanibekov returned to earth aboard Soyuz T-13 with Soyuz T-14 crewman Georgi Grechko of the Soyuz T-14 crew, leaving the Soyuz T-14 spacecraft and the rest of its crew aboard Salyut 7 through November 21.

Below: Discovery **51-G lifts off. This was the first space flight in history to carry personnel from three countries (see text).** *At left:* **Astronaut Bill Fisher rides the** *Discovery* **51-I's Remote Manipulator Arm during the repair and rescue of the crippled Leasat that had been launched by STS 51-D.**

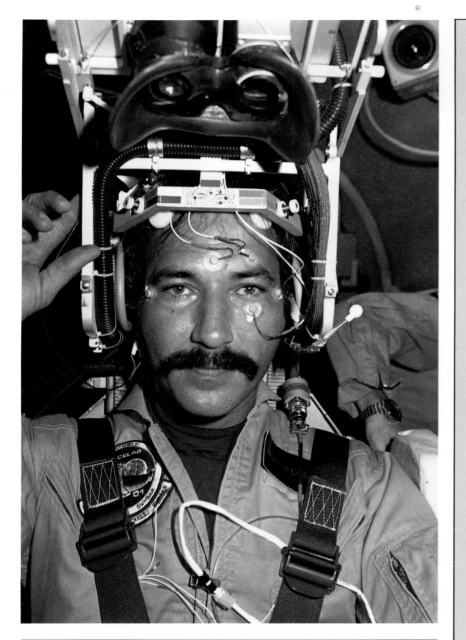

At top left: Challenger 61-A payload specialist Wubbo Ockels, of the Netherlands. The 61-A mission also had among its record eight-member crew West German payload specialists Ernst Messerschmid and Reinhard Furrer.

This mission was also known as the Spacelab D-1 mission, for its ESA Spacelab Experimentation Module (D-1 for its West German experimental slant) payload. Control of Spacelab was given, for part of the mission, to the German Space Operations Center—the first time that operational control of a US space flight had been given to an entity outside of the United States.

At left: The French National Space Agency's (CNES) Patrick Baudry, who flew aboard *Discovery* 51-G. *Above:* Salyut 7 cosmonauts (left to right) Viktor Savinykh, Vladimir Vasyutin and Alexander Volkov. Savinykh boarded Salyut 7 from Soyuz T-13, and Vasyutin and Volkov, of Soyuz T-14, joined him. *At right:* Supplying (space tug) and manning (Soyuz) Salyut 7.

Cosmos 1686 module with solar panels extended

Salyut space station

Soyuz spacecraft used by resident or host Salyut crew

At left: The European Space Agency's comet probe Giotto was launched in 1985 to intersect Halley's comet on its approach to Earth. *Above:* Japan's Planet-A Halley's Comet probe. *Below:* One of three Shuttle-launched AUSSAT satellites. The AUSSAT series represents Australia's first national communications satellites.

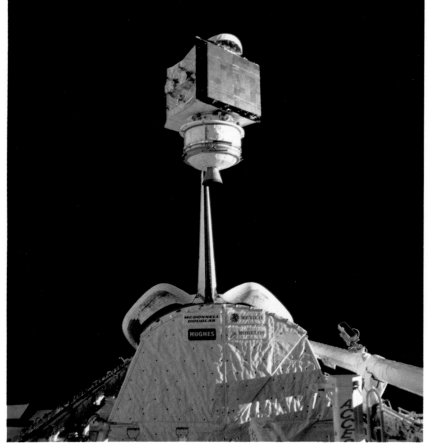

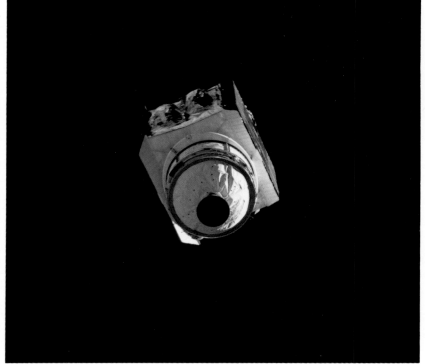

At top of page: Mexico's Morelos-B communications satellite is here launched out of its cradle during *Atlantis* Shuttle Orbiter mission 61-B. *At left:* The RCA SATCOM K-2 communications satellite is here spun out of its cradle (behind that of the Morelos satellite) aboard *Atlantis* 61-B, and *(above)* quickly moves away from the Space Shuttle and into orbit.

Space Flight Highlights of 1985

	Launch date	Launch vehicle	Launch weight
DSPS (USA)	25 Jan	Shuttle	2000 lb
Soyuz T-13 (USSR)	6 Jun	A-2	15,400 lb
Giotto (ESA)	2 Jul	Ariane	2112 lb
Planet A 'Suisei' (Japan)	19 Aug	M-3SII	275 lb
AUSSAT (Australia)	28 Aug	Shuttle	1442 lb
Leasat 4 (USA)	28 Aug	Shuttle	2900 lb
Soyuz T-14 (USSR)	17 Sep	A-2	15,400 lb
Morelos B (Mexico)	27 Nov	Shuttle	2000 lb
Satcom K-2 (USA)	28 Nov	Shuttle	4144 lb

American Space Shuttle Missions
Challenger (OV-99)

1.	29 Apr	(51-B/ Spacelab 3)	Frederick Gregory
			Don Lind
			Robert Overmyer
			Norman Thagard
			William Thornton
			Taylor Wang
			Lodewijk van Den Berg
2.	29 Jul	(51-F/ Spacelab 2)	Loren Acton
			John Bartol
			Anthony England
			Roy Bridges, Jr
			Gordon Fullerton
			Karl Henize
			Story Musgrave
3.	30 Oct	(61-A/ Spacelab D-1)	Guion Bluford, Jr
			James Buchli
			Bonnie Dunbar
			Henry Hartsfield, Jr
			Steven Nagel

American Space Shuttle Missions (continued)
Discovery (OV-103)

1.	24 Jan	(51-C)	James Buchli
			Thomas Mattingly
			Ellison Onizuka
			Gary Payton
			Loren Shriver
2.	12 Apr	(51-D)	Karol Bobko
			E J 'Jake' Garn
			David Griggs
			Jeffrey Hoffman
			Rhea Seddon
			Charles Walker
			Donald Williams
3.	17 Jun	(51-G)	Daniel Brandenstein
			John Creighton
			John Fabian
			Shannon Lucid
			Steven Nagel
4.	27 Aug	(51-I)	Richard Covey
			Joe Engle
			William Fisher
			Michael Lounge
			James van Hoften

Atlantis (OV-104)

1.	3 Oct	(51-J)	Karol Bobko
			Ronald Grabe
			David Hilmers
			William Pailes
			Robert Stewart
2.	27 Nov	(61-B)	Mary Cleave
			Bryan O' Connor
			Jerry Ross
			Brewster Shaw, Jr
			Sherwood Spring
			Charles Walker

Soviet Cosmonauts

44.	Vladimir Dzhanibekov (5th flight)	(Soyuz T-13)	6 Jun
51.	Victor Savinykh (2nd flight)		
60.	Vladimir Vasyutin	(Soyuz T-14)	17 Sep
34.	Georgi Grechko (3rd flight)		
61.	Alexander Volkov		

Foreign Payload Specialists on American Space Shuttle Missions

1.	Patrick Baudry (France)	(51-G)	17 Jun
2.	Salman Abdel-aziz Al Saud (Saudi Arabia)	(51-G)	17 Jun
3.	Reinhard Furrer (West Germany)	(61-A/ Spacelab D1)	30 Oct
4.	Ernst Messerschmid (West Germany)	(61-A/ Spacelab D1)	30 Oct
5.	Wubbo Ockels (Netherlands)	(61-A/ Spacelab D1)	30 Oct
6.	Rodolfo Neri (Mexico)	(61-B)	26 Nov

Above: The USAF/TRW (Thompson Ramo Wooldridge, Inc) Defense Support Program Satellite (DSPS) is shown following a pre-launch inspection. This satellite was launched by *Discovery* on the semi-classified 51-C mission in January of 1985.

1977 Flight Tests   STS-1 (1981)   STS-2 (1981)   STS-3 (1982)   STS-4 (1982)   STS-5 (1981)

41-C (1984)   41-D (1984)   41-G (1984)   51-A (1984)   51-C (1985)

The Space Shuttle Transportation System 1977–1986

In 1977, the US began its successful atmospheric glide tests of the Shuttle Orbiter prototype *Enterprise,* but the *Space* Shuttle project suffered delays—a Shuttle Orbiter was to enter orbit sometime in 1979, but did not do so until, on 12 April 1981, Astronauts John Young and Robert Crippen took off in the Shuttle Orbiter *Columbia* and guided their craft through a 54-hour orbital mission to a perfect ground landing.

Columbia returned to space on 12 November, and flew three times in 1982. In 1983, Sally Ride became the first American woman in space, Guion Bluford became the first black American man in space and West Germany's Ulf Merbold became the first non-American to fly aboard an American spacecraft. The latter mission also saw the first use of the ESA Spacelab Module. The first three missions of 1983 were flown aboard the second Shuttle Orbiter, the freshly inaugurated *Challenger.*

Challenger was followed by the inaugurations of two more Shuttle Orbiters—*Discovery* in 1984 and *Atlantis* in 1985. The reusable, and heavily used, Shuttle Orbiters were employed for scientific tests; launched, rescued and repaired satellites, and set the record for manned space flights in a single year at nine in 1985. The year 1986, as is discussed in the following chapter, brought tragedy to the Shuttle Orbiter program.

After the *Challenger/* 51-L tragedy, several years were consumed in testing and re-testing the STS equipment and systems for safety. At this point, general plans are in the works for a Shuttle mission sometime near the end of the present decade.

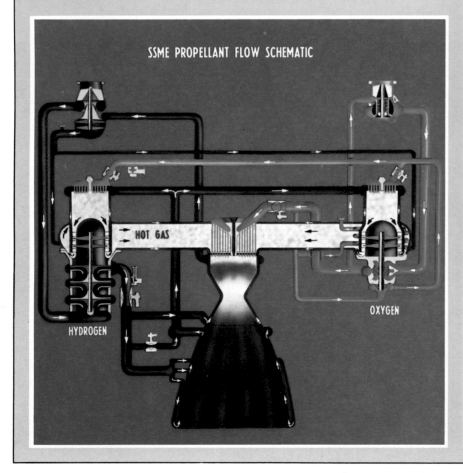

SSME PROPELLANT FLOW SCHEMATIC

HOT GAS

HYDROGEN

OXYGEN

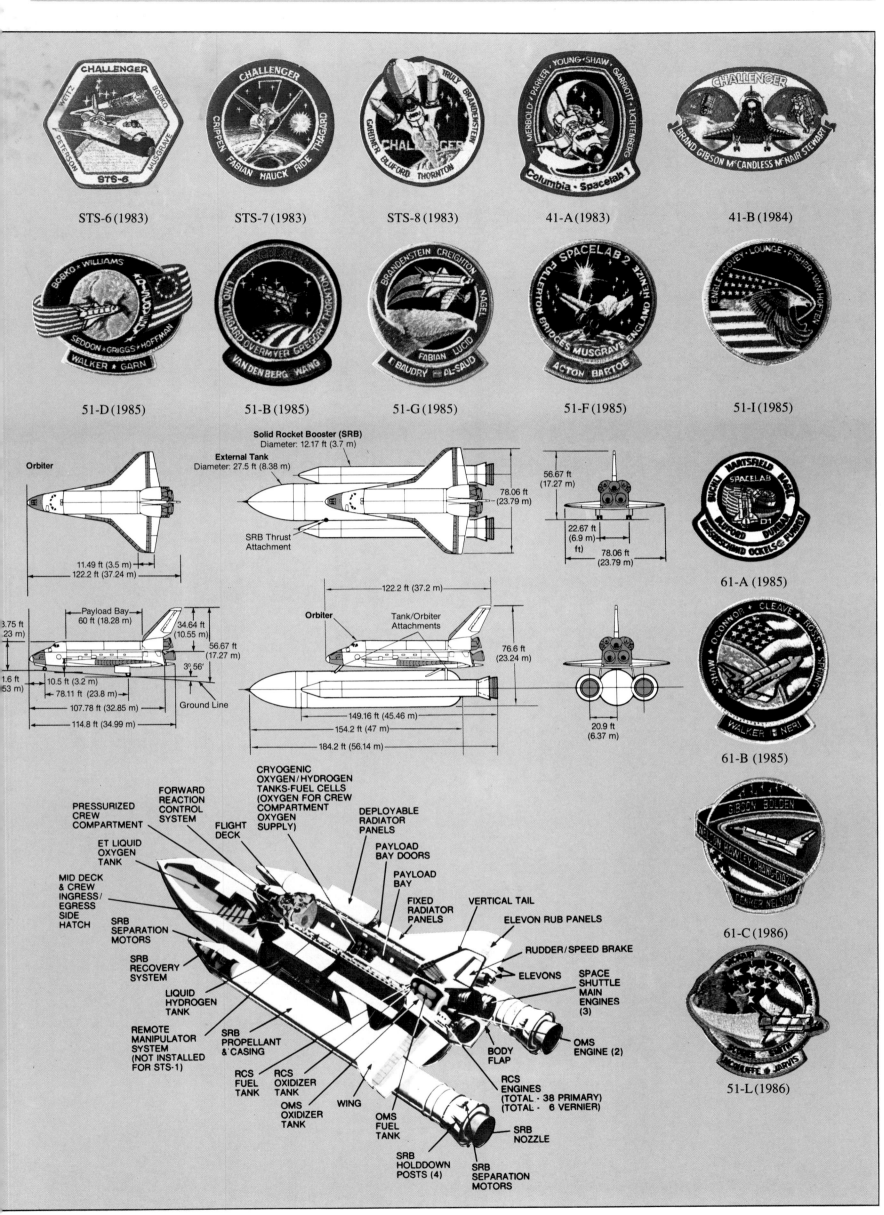

STS-6 (1983)

STS-7 (1983)

STS-8 (1983)

41-A (1983)

41-B (1984)

51-D (1985)

51-B (1985)

51-G (1985)

51-F (1985)

51-I (1985)

61-A (1985)

61-B (1985)

61-C (1986)

51-L (1986)

Orbiter

Solid Rocket Booster (SRB)
Diameter: 12.17 ft (3.7 m)

External Tank
Diameter: 27.5 ft (8.38 m)

SRB Thrust Attachment

78.06 ft (23.79 m)

56.67 ft (17.27 m)

22.67 ft (6.9 m) ft)

78.06 ft (23.79 m)

11.49 ft (3.5 m)

122.2 ft (37.24 m)

3.75 ft 23 m)

1.6 ft 53 m)

Payload Bay 60 ft (18.28 m)

34.64 ft (10.55 m)

56.67 ft (17.27 m)

3° 56'

10.5 ft (3.2 m)

78.11 ft (23.8 m)

107.78 ft (32.85 m)

114.8 ft (34.99 m)

Ground Line

122.2 ft (37.2 m)

Orbiter

Tank/Orbiter Attachments

76.6 ft (23.24 m)

20.9 ft (6.37 m)

149.16 ft (45.46 m)

154.2 ft (47 m)

184.2 ft (56.14 m)

PRESSURIZED CREW COMPARTMENT

FORWARD REACTION CONTROL SYSTEM

FLIGHT DECK

CRYOGENIC OXYGEN/HYDROGEN TANKS-FUEL CELLS (OXYGEN FOR CREW COMPARTMENT OXYGEN SUPPLY)

DEPLOYABLE RADIATOR PANELS

ET LIQUID OXYGEN TANK

PAYLOAD BAY DOORS

PAYLOAD BAY

MID DECK & CREW INGRESS/ EGRESS SIDE HATCH

FIXED RADIATOR PANELS

VERTICAL TAIL

ELEVON RUB PANELS

SRB SEPARATION MOTORS

RUDDER/SPEED BRAKE

SRB RECOVERY SYSTEM

ELEVONS

SPACE SHUTTLE MAIN ENGINES (3)

LIQUID HYDROGEN TANK

REMOTE MANIPULATOR SYSTEM (NOT INSTALLED FOR STS-1)

SRB PROPELLANT & CASING

BODY FLAP

OMS ENGINE (2)

RCS FUEL TANK

RCS OXIDIZER TANK

RCS ENGINES (TOTAL - 38 PRIMARY) (TOTAL - 6 VERNIER)

OMS OXIDIZER TANK

WING

OMS FUEL TANK

SRB NOZZLE

SRB HOLDDOWN POSTS (4)

SRB SEPARATION MOTORS

1986

I t promised to be a banner year in space for the United States. All four Space Shuttle Orbiters were functional and had at least two missions under their belts. (*Challenger* had *nine*!) After not having made a flight since November 1983, the venerable *Columbia* had gotten back into the swing by making the first flight of 1986, a six-day mission launched on 12 January.

Columbia's 1986 flight was, in fact, a 1985 scheduled mission, so the STS docket still had 15 flights on it. They were to include the launch of the Hubble Space Telescope, the biggest telescope ever used in space and expected to be more powerful than anything on earth due to the absence of atmospheric interference in space. Other STS payloads would include the Galileo Jupiter Mission that would actually send probes into the Jovian atmosphere.

Everything was shaping up for a good year for NASA. Halley's Comet would be observed by human beings in space, the incredible Voyager 2 spacecraft was flawlessly making its way toward the first ever encounter of a spacecraft with the planet Uranus. The Space Shuttle had even become so routine that a high school teacher would be flying on Mission 51-L in January!

Voyager 2 came through right on cue for the United States with a textbook-perfect flyby of Uranus on the weekend of 24 January. The murky blue atmosphere of the planet and the icy crevices of her five largest moons were revealed in spectacular detail. There were even ten newly discovered moons to add to the Atlas of the Solar System!

The following week, on 27 January, *Challenger* thundered aloft from Kennedy Space Center for the 10th and final time. Unknown to anyone at that moment, flames were already licking through a failed seal in one of the solid rocket boosters that propelled *Challenger* upward. Swiftly they grew into a sheet of flames, and 73 seconds after launch the liquid oxygen and liquid hydrogen in the big rust-red fuel tank ignited in a gargantuan fireball that destroyed both the tank and *Challenger*. The two solid rocket boosters twisted across the sky like blinded, frightened animals. Seven Americans—including Concord, New Hampshire high school teacher Sharon Christa McAuliffe—perished in the disaster.

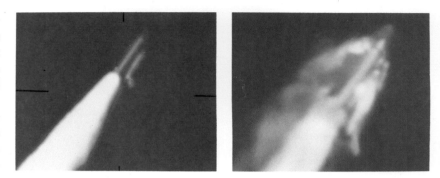

Above left: A faint flame, above the rocket exhausts in this *Challenger/* 51-L takeoff photo, erupts into an explosion at *above right. Below:* The *Challenger/* 51-L crew, left to right, front to back: Michael J Smith, Francis R Scobee, Ronald E McNair, Ellison S Onizuka, Sharon Christa McAuliffe, Gregory Jarvis and Judith A Resnik. *At right:* *Challenger* Teacher in Space Program citizen payload specialist Sharon Christa McAuliffe, the social studies teacher who was to have taught her class from orbit.

NASA announced that there would not now be 15 STS missions in 1986. The next mission was postponed six months, postponed again, and then postponed indefinitely. The year 1986 would end with no further Space Shuttle launches and, though no one dared to predict it, there would be no launches in 1987 either.

Though the United States manned space program would be completely derailed for over two years, the *Challenger* disaster should be put into perspective. Seven people had died a terrible death, but in a quarter century of manned space flight, with nearly 100,000 hours of space time logged, the total death toll stood at only 14, a small number considering the risks involved. Of these 12 men and two women, three Americans had died aboard a nonflying spacecraft on the ground (Apollo 1, 1967), four Russians had died returning to earth (Soyuz 1, 1967 and Soyuz 11, 1971), and now seven Americans had died 73 seconds after launch. Technically, there had yet to be a single fatality during operations in outer space or on the moon.

continued on page 197

Space Flight
Highlights of 1986

	Launch date	Launch vehicle	Launch weight
Mir space station (USSR)	19 Feb	D-1	56,600 lb
Spot 1 (France)	22 Feb	Ariane	1540 lb
Soyuz T-15 (USSR)	13 Mar	A-2	15,400 lb
FLTSSATCOM (USA)	5 Dec	Atlas/Centaur	2300 lb

Soviet Cosmonauts
49.	Leonoid Kizim (3rd flight)	(Soyuz T-15)	13 Mar
57.	Vladimir Solovyev (2nd flight)		

American Space Shuttle Missions
Columbia (OV-102)
1.	12 Jan	(61-C)

Charles Bolden, Jr.
Robert Cenkar
Franklin Chang-Diaz
Robert Gibson
Steven Hawley
Bill Nelson
George Nelson

Challenger (OV-99)
1.	28 Jan	(51-L)

Gregory Jarvis
Christa McAuliffe
Ronald McNair
Ellison Onizuka
Judy Resnik
Francis Scobee
Mike Smith

Grounded! *Below:* **The Space Shuttle *Atlantis* is rolled out to Kennedy Space Center's Pad 39-B in October of 1986. The *Atlantis* stayed put for seven weeks, serving as the base for numerous emergency contingency procedure tests, and then was rolled back** into storage. *At left:* **A French-built Ariane rocket blasts off on a European Space Agency mission; with much intense concentration on the use and development of the Space Shuttle program, the US had let its own fleet of expendable boosters diminish.**

The Soviet Mir Space Station Complex

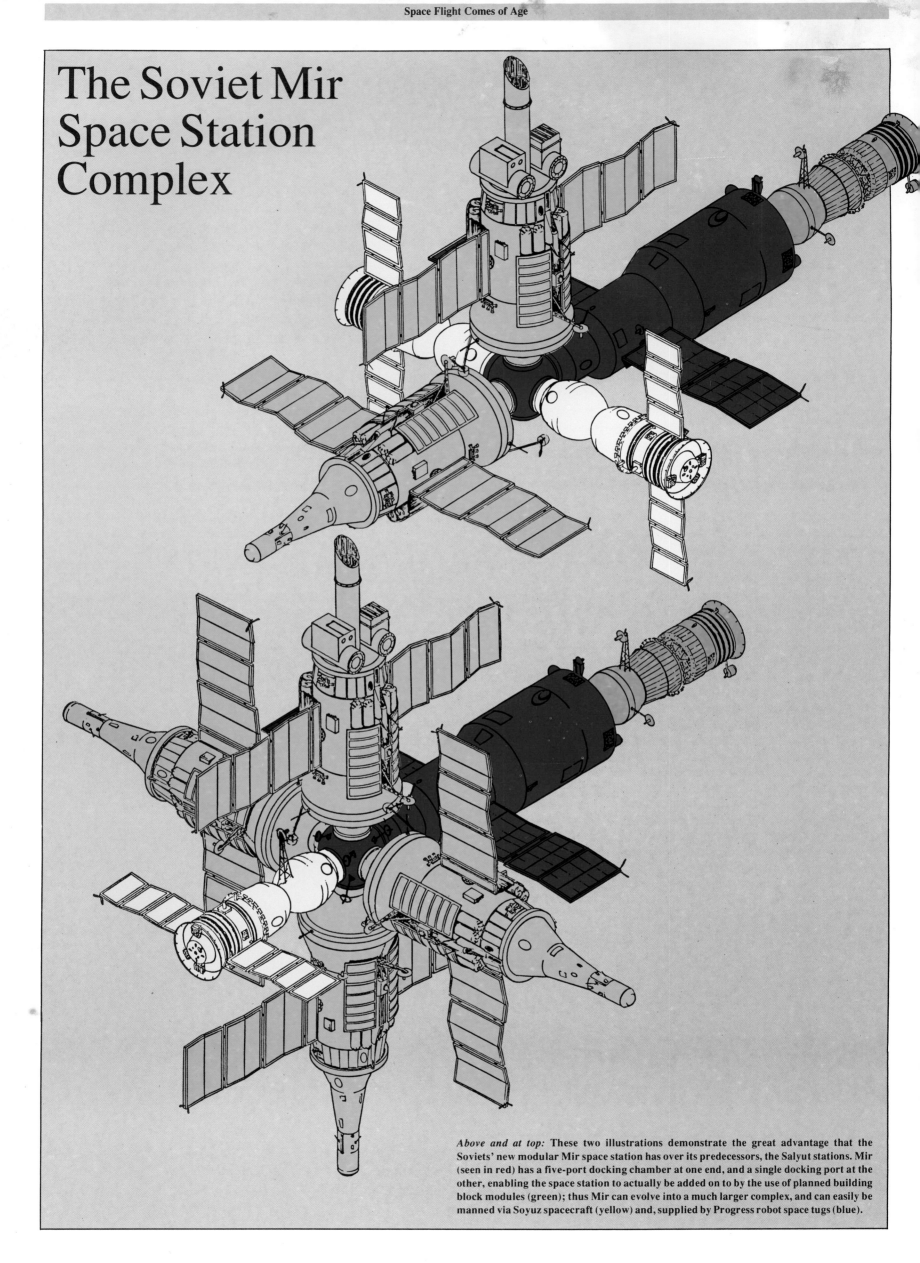

Above and at top: These two illustrations demonstrate the great advantage that the Soviets' new modular Mir space station has over its predecessors, the Salyut stations. Mir (seen in red) has a five-port docking chamber at one end, and a single docking port at the other, enabling the space station to actually be added on to by the use of planned building block modules (green); thus Mir can evolve into a much larger complex, and can easily be manned via Soyuz spacecraft (yellow) and, supplied by Progress robot space tugs (blue).

continued from page 192

Having lost their own cosmonauts, the *Challenger* disaster must have been a sobering sight for the Soviets as well. The fact that the United States was canceling the Space Shuttle launches went beyond the issue of manned space flight alone. For several years the United States had also looked upon the Space Shuttle as a means of launching satellites to such a degree that it had all but abandoned the use of expendable launch vehicles (ELV)—or one-time-use rockets—for putting satellites in orbit. Theoretically, with a fleet of Shuttle Orbiters that could launch three or four satellites per mission, there was little need for old fashioned, nonreusable spacecraft. However, with the STS system now grounded there were suddenly too many payloads to launch and too few ELVs to do it. This fact handed the Soviet Union the leadership role in space on a platter. The record for 1985 speaks for itself. The Americans had launched nine missions of the reusable shuttle, while the Soviets had launched nothing comparable, *but* the Soviet Union had launched 98 ELVs while the United States had launched just *eight*! Furthermore, the Soviets had launched five multiple-payload missions for a total of 119 payloads in space, and their Cosmos all-purpose, unmanned program had seen the launch of the 1700th spacecraft on 25 October 1985.

On 19 February, the Soviets launched Mir, the first of a new class of space stations that are similar to, but larger than, the Salyut series. On 13 March, Leonid Kizim and Vladimir Solovyev were launched aboard Soyuz T-15 to become the first men to staff the new space station.

In the meantime, an armada of international spacecraft (with the United States conspicuously absent), were closing in for the first space-based observations of Halley's Comet. The Soviet Vega 1 came within 5520 miles of Halley's on 6 March, while Vega 2 made a high speed flyby on 9 March, coming to within 4987 miles and returning the first images ever seen of the famous comet's icy core. Next in line was the European Space Agency's Giotto spacecraft, which had been launched by a French Ariane rocket in July 1985 from the launch center in French Guiana. Equipped with a multicolor camera developed by the Max Planck Institut fur Aeronomie in West Germany, Giotto made its closest approach to Halley's Comet, a scant 366 miles away, on 14 March.

By mid-April the Soviet Union had launched 11 reconnaissance satellites and 12 other military navigation and communication spacecraft in 1986, while the United States had launched only one. On 18 April, a Big Bird military reconnaissance satellite was launched from Vandenberg AFB, California. Just 8.5 seconds after launch, the Titan 34D ELV carrying the spacecraft exploded in a shower of fire and debris over the Pacific. The loss of this mission helped to thicken the gloom surrounding the wake of the *Challenger* disaster and to underscore the shortage of heavy lift ELVs, as Titan class launch vehicles were now grounded for six to 12 months.

On 6 May, Leonid Kizim and Vladimir Solovyev, who had been aboard the Mir space station for almost two months, transferred to the Salyut 7 space station using their Soyuz T-15 spacecraft. On 21 May, the Soviets unveiled the first of the new Soyuz TM spacecraft when the unmanned prototype was launched into space for a 23 May robot rendezvous with Mir.

Even as the Soyuz TM was docking with Mir, Kizim and Solovyev were conducting an almost four-hour spacewalk and Soviet crews were launching a trio of unmanned satellites in rapid succession. These included Cosmos 1745 (a military payload) on 23 May, Ekran (a TV broadcasting satellite) the next day, and a Meteor 2 weather satellite on 27 May.

On 26 June the Soyuz T-15 cosmonauts returned to Mir and on 3 July Leonid Kizim completed 362 cumulative days in space, breaking Valery Ryumin's old record established in 1981. Three days later he became the first man to have logged a year's total space time. On 16 July, Mir's first crew and the only cosmonauts to go into space in 1986 came home aboard Soyuz T-15 after 125 days.

Below: This photo of the icy core of Halley's Comet was taken by the European Space Agency comet probe, Giotto, from very close in. Note Halley's gas envelope, and the gas jets shooting out of the comet's core.

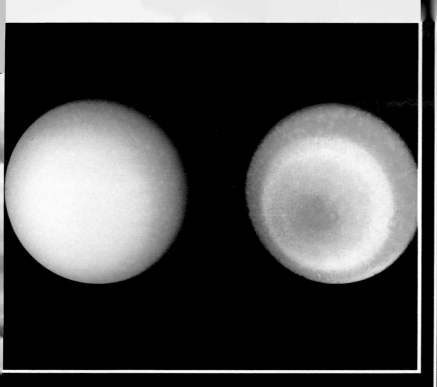

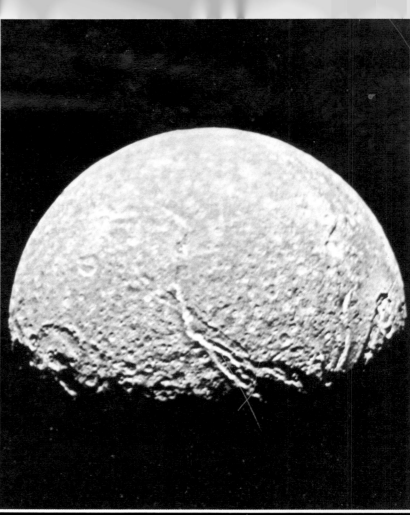

Above: These Voyager 2 images show the giant gaseous planet Uranus in real color (left) and false color (right). False color is often used to highlight a particular planetary, lunar or solar feature. *At right:* The heavily-featured Uranian moon Titania, which has nothing in the way of folds, faults or trenches over Uranian moon Miranda, which is shown *below*—with Elsinore Corona in evidence at left and Inverness Corona at right—and is shown at *below right*—with Inverness Corona to the right. *Opposite:* The huge, gaseous Uranus shows a blue crescent to the departing Voyager 2's cameras. Voyager 2 will continue transmitting into the 21st century, as will Voyager 1. Both space probes will eventually transmit data from beyond the known Solar System.

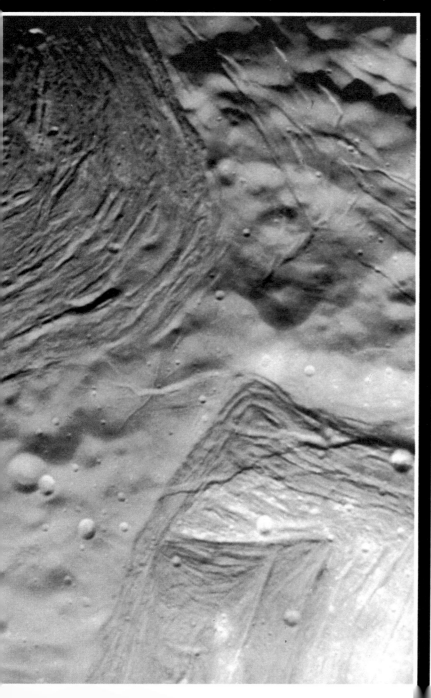

Spacecraft
in Service

Appendix: Spacecraft in Service

Arab Satellite Communications Organization (ASCO)

Major Spacecraft Classes in Service, 1987

	Typical launch vehicle	Total number launched	Year of first launch
Arabsat (Communications)	Ariane/ Space Shuttle	2	1985

Principal Launch Site: None (Spacecraft launched by NASA or ESA)

Australia

Major Spacecraft Classes in Service, 1987

	Typical launch vehicle	Total number launched	Year of first launch
Aussat (Communications)	Ariane/ Space Shuttle	3	1985

Principal Launch Site: None (Spacecraft launched by NASA or ESA)

Brazil

Major Spacecraft Classes in Service, 1987

	Typical launch vehicle	Total number launched	Year of first launch
SBTS (Communications)	Ariane	2	1985

Principal Launch Site: None (Spacecraft are launched by ESA)

Canada (Telesat)

Major Spacecraft Classes in Service, 1987

	Typical launch vehicle	Total number launched	Year of first launch
Anik D (Communications)	Space Shuttle or Delta	2	1982

Principal Launch Site: None (Spacecraft are launched by NASA)

China

Major Spacecraft Classes in Service, 1987

	Typical launch vehicle	Total number launched	Year of first launch
'China' series (Communications)	Long March (C2)	18	1970*
Fengyun (Weather)	Long March (C2)	1	1987

Principal Launch Site: Xichang (near Chengdu), China

*Current spacecraft includes China 18 (1986)

European Space Agency (ESA)

Major Spacecraft Classes in Service, 1987

	Typical launch vehicle	Total number launched	Year of first launch
ECS (Communications)	Ariane	3*	1983
Exosat (X-ray observatory)	Delta	1	1983
Meteosat (Weather)	Delta	2	1977

Principal Launch Site: Kourou, French Guiana

*launch of ECS-3 failed

Federal Republic of Germany

Major Spacecraft Classes in Service, 1987

	Typical launch vehicle	Total number launched	Year of first launch
SPAS (Free flying photographic, used in conjunction with Space Shuttle)	Space Shuttle	2	1983

Principal Launch Site: None (German SPAS spacecraft are launched only from the NASA Space Shuttle, but German aerospace firms are also representd in most ESA projects)

France (Centre National d'Etudes Spatiales, CNES)

Major Spacecraft Classes in Service, 1987

	Typical launch vehicle	Total number launched	Year of first launch
SPOT (High-resolution photographic)	Ariane	1	1986
Telecom (Communications)	Ariane	3	1984

Principal Launch Site: Kourou, French Guiana

India (Indian Space Research Organization, ISRO)

Major Spacecraft Classes in Service, 1987

	Typical launch vehicle	Total number launched	Year of first launch
Bhaskara (Earth observation)	C-1	2	1977
Insat	Delta	4	1982

Principal Launch Site: (Most Indian spacecraft are launched by the USSR or NASA)

Indonesia

Major Spacecraft Classes in Service, 1987

	Typical launch vehicle	Total number launched	Year of first launch
Palapa (Domestic communications)	Space Shuttle	3	1976

Principal Launch Site: None (Indonesian spacecraft are launched by NASA)

Japan (National Space Development Agency, NASDA)

Major Spacecraft Classes in Service, 1987

	Typical launch vehicle	Total number launched	Year of first launch
ASTRO (Astrophysics)*	M-3S, N-3S	2	1983 *
BS-2 (Broadcasting)	N-II	2	1984
EGS (Geosurvey)	H-1	1	1986
GMS (Weather)	Delta/N-II	3	1977
Planet-A (Venus/ Halley's Comet probe)*	M-3S II	1	1985 *

Principal Launch Site: Tanegashima Space Center, Japan

*These are Institute of Space and Astronautical Science (ISAS) satellites, not NASDA.

Mexico

Major Spacecraft Classes in Service, 1987

	Typical launch vehicle	Total number launched	Year of first launch
Morelos	Space Shuttle	2	1985

Principal Launch Site: None (Mexican spacecraft are launched by NASA)

Soviet Union

Major Spacecraft Classes in Service, 1987

	Typical launch vehicle	Total number launched	Year of first launch
Cosmos (Multipurpose)	SL-12	1887*	1962
Meteor (Weather)***	SL-4	42***	1975***
Molniya (Communications)	SL-6	110	1965
Progress (Space station supply ship)	SL-4	27**	1978
Soyuz (Manned)	SL-4	40	1967
Soyuz T (Manned)	SL-4	15	1980
Soyuz TM (Manned)	SL-4	3	1986

Principal Launch Site: Baikanur Cosmodrome (Tyurantam, USSR) and Northern Cosmodrome (Plesetsk, USSR)

*As of October 1987
**As of February 1987
***Includes Meteor 1 series (27 launches)

Sweden

Major Spacecraft Classes in Service, 1987

	Typical launch vehicle	Total number launched	Year of first launch
Viking (Research)	Ariane	1	1986

Principal Launch Site: None (Swedish spacecraft are launched through ESA)

United Kingdom

Major Spacecraft Classes in Service, 1987

	Typical launch vehicle	Total number launched	Year of first launch
Skynet series* (Communications)	Ariane/or Space Shuttle	3	1969

Principal Launch Site: None (UK spacecraft were first launched by Australia and later by NASA, but given the UK's participation in ESA, most future UK launches will probably occur under ESA auspices)

*Skynet 1 was the first UK military satellite. No Skynet 4 has yet been launched, but three are on order for the UK and two for NATO. Skynet 2 series consisted of two satelllites. There was no Skynet 3.

United States (Commercial)

Major Communications Spacecraft Classes in Service, 1987

	Typical launch vehicle	Total number launched	Year of first launch
American Satellite	Space Shuttle	1	1985
Comstar	Atlas/Centaur	4	1976
G Star	Ariane	3	1985
Galaxy	Delta	3	1983
Marisat	Delta	3	1976
SBS	Delta/ Space Shuttle/or Ariane	5	1980
Spacenet	Ariane	3	1985
Telstar	Delta	5	1962*
Westar	Delta/or Space Shuttle	7	1974

Principal Launch Site: None (American commercial spacecraft were launched only by NASA until 1984, when American users started placing payloads with ESA as well)

*The first launch of current Telstar II series was in 1983

United States (Military)

Major Spacecraft Classes in Service, 1987

	Typical launch vehicle	Total number launched	Year of first launch
Big Bird (Reconnaissance)	Titan/or Space Shuttle	*	1971
DMSP (Weather)	Atlas	*	1971
DSCS (Communications)	Titan	42+	1966
FLTSATCOM (Communications)	Atlas/Centaur	8	1973
Keyhole (Reconnaissance)	Titan/or Space Shuttle	*	1976
Leasat (Communications)	Space Shuttle	1	1984
Navstar** (Navigation)	Atlas	6	1978

Principal Launch Site: Vandenberg AFB, California

*Classified
**aka Global Positioning System

United States (National Aeronautics & Space Administration, NASA)

Major Spacecraft Classes in Service, 1987

	Typical launch vehicle	Total number launched	Year of first launch
GOES (Weather)	Delta/or Space Shuttle	11	1975
Explorer*	Delta	65	1958
Landsat (Earth resources)	Delta	5	1972
NOAA (Weather)	Atlas	16	1970
Pioneer 10/11 (Deep space exploration)	Atlas/Centaur	2	1972
Space Shuttle Transportation System	(Self-contained)	**	1981
Voyager (Outer solar system exploration)	Titan/Centaur	2	1977

Principal Launch Site: Kennedy Space Center, Cape Canaveral, Florida

*Current spacecraft include Dynamics Explorer (1981) and Solar Mesosphere Explorer (1981)
**Four reusable spacecraft were used 24 times until the system was grounded in 1986

Index

Abbey, George: *142*
Acton, Loren: 189
Agena: 38, *39*
Aksynov, Vladimir: 99,135
Aldrin, Edwin, 38, 53, 54
Alexandrov, Alexander: 158, 165, 200
Allen, Joseph P IV: *152–153, 154,* 155, 177, *178–179*
Al-Saud, Prince Salman Abdel-Aziz: 182, 189
American Satellite (United States): 205
Anders, William: 46, 49
Anik-C (Canada): 150, *150,*155
Anik-C2 (Canada): *163*, 165, 169
Anik-D (Canada): 204
Apollo: 32, 38, 84, *86*
Apollo 1: 42, *42*, 192
Apollo 7: 46, *46,*49,*54*
Apollo 8: 46, *47,* 49, 50, 64
Apollo 9: 50, *50, 51,* 53
Apollo 10: 50, *50,* 53, 84
Apollo 11: *52,* 53, 54
Apollo 12: 53, *54*
Apollo 13: 56, 58, *68*
Apollo 14: 56, 58, 61
Apollo 15: 58, *59,* 61
Apollo 16: 64, *64,* 67
Apollo 17: 64, *64,* 67
Apollo/Skylab 1: 75
Apollo/Skylab 2: 75
Apollo/Skylab 3: 75
Apollo-Soyuz Test Project (ASTP): *84, 87*
Argyre Planitia: *105*
Ariane: 197, 200
Arabsat (ASCO): 204
Armstrong, Neil: 38, 53, 54, *54*
Arsia Mons: *104*
Artyukhin, Yuri: 78
Ascraeus Mons: *104*
A-1 Asterix (France): 32, 37
ASTRO (Japan): 204
Astronaut Group 1: *18*
Atkov, Oleg: 170, 177
Atlantis (OV-104): 182, 189,190, *195,*
AUSSAT: (Australia): *187,* 189, 204
Bartol, John: 189
Baudry, Patrick: 182, *184,* 189
Bean, Alan: 53, 54, 75
Belyayev, Pavel: 32, 37
Beregovoi, Giorgi: 32, 49
Berezovio, Anatoli: 155
Bhaskara (India): 204
Biblis Patera: *104*
Big Bird (US Military): 205
Big Joe (Mercury): *102*
Blagov, Viktor: 200
Bluford, Guion: 158, 161, 165, 189, 190
Bobko, Karol: 165, 189
Bolden, Charles, Jr: 195
Borman, Frank: 32, 37, 46, 49
Buchli, James: 189
Brand, Vance D: 84, 93, *152-153,* 155, 177
Brandenstein, Daniel: 165, 189
Bridges, Roy Jr: 189
BS-2 (Japan): 205
Bykovsky, Valery: 24, 27, 99, 114, 119
Carpenter, Scott: *18*,20, *20,* 23
Carr, Gerald: 75, 77, 78
Cenkar, Robert: 195
Centre National D'Etudes Spatiales (CNES): 32
Cernan, Gene: 38, *39,* 50, 53, 64, 67
Chaffee, Roger: 42, *42*
Challenger (OV-99): 158, *165, 168-169,* 177, 182, 189, 190, 192, 195
Chang-Diaz, Franklin: 195
'China' Series (China): 56, 204
Chreighton, John: 189
Chretien, Jean-Loup: 150, 155, *155*
Chryse Planitia: *104*
Cleave, Mary: 189
Coats, Michael L: *171,* 177
Collins, Michael: 38, 53, 54
Columbia (OV-102): 140, *140,* 143, 144, 150, 158, 182, 190, 192, 195
COMSTAR: *97,* 99 *101,* 205
Conrad, Charles: 32, 37, 38, 53, 54, 75
Cooper, Gordon: *18,* 24, 27, 32, 37
Cosmos Series Spacecraft (USSR): 23, 24, 42, 124, 127, 188, 197, 200, 205
Covey, Richard: 189
Crippen, Robert: 140, *142-143,* 143, 165, 177, 190
Cunningham, Walter: 46, 49, *49*

D-Series Launch Vehicles (Soviet): 32, 42
Delta: *201*
Demin, Lev: 78
Dione: *136, 148-149*
Discovery (OV-103): 170, *170,* 177, 182, 189, 190
DMSP (US Military): 205
Dobrovolsky, Georgi: 58
DSCS 1, 2, and 3: 151, *150–151,* 155, 205 (US Military)
DSPS: *189*
Duke, Charles: 64, 67
Dunbar, Bonnnie: 189
Dzhanibekov, Vladimir: 119, 140, 143, 155, 177, 182, 189
Earth Observing System: 200
EC 3 (European Space Agency): 204
Echo 1: 14, *15,* 15
EGS (Japan): 205
Eisele, Donn: 46, 49, *49*
Ekram: 197
Enceladus: *148*
England, Anthony: 189
Engle, Joe: 106, 140, *142,* 143, 189
Enterprise (OV-101): *106, 112,* 190
ESSA 7 (TOS-E) (USA): 49
Evans, Ronald: 67
EXOSAT (European Space Agency): 204
Expendable Launch Vehicles (ELV): 197
Explorer (NASA): 205
Explorer 1: 10, *11*
Explorer 2: 10
Explorer 3: *11*
Explorer 4: 10
Explorer 6: 10
Explorer 18: *27*
Explorer 19: 24
Explorer 20: *2–3*
Explorer 29: 94
Explorer 35: 44
Explorer 36: 42, 94
Explorer 47: 67
Explorer 48: 67
Fabian, John: 165, 189
Faris, Mohammed: 200
Farkas, Bertalan: 130, 135, *135*
Fengyun (China): 204
Feoktistov, Konstantin: *28,* 31
Filipchenko, Anatoli: 53, 78
Fisher, Anna: 177
Fisher, Bill: *183,* 189
FLTSATCOM: *126–127,* 127, 195, 205
Fullerton, Gordon: 106, 155, 189
Furrer, Reinhard: *184,* 189
G Star (United States): 205
Gagarin, Yuri: 18, *19,* 38
Galaxy 4 (USA): 165, 205
Gardner, Dale: 165, 177, *178–179*
Garn, 'Jake' EF: 189
Garneau, Marc: 170, *176,* 177
Garriott, Owen: 75, 165
Gemini Program: 32, *33, 36, 37,* 38
Gemini 3: 37
Gemini 4: 37
Gemini 5: 32, 37
Gemini 6: 32, 37, 84
Gemini 7: 32, 37
Gemini 8: 38
Gemini 9: 38, 84
Gemini 10: 38
Gemini 11: 38
Gemini 12: 38
GEOS A, B, and C: 93, 94, *94*
GEOS/ESSA: *117,* 119
Geostationary Meterological Satellite (GMS): *113,* 177
Giotto: *186–187,* 189, 197
GOES (NASA): 205
GOES A: 93, *93*
GOES C. D. E, F and H: *93,*119, *120, 132,* 135 121, *121,* 200, *201*
Gibson, Edward: 75, 77, 78
Gibson, Robert: 177, 195
Glazkov, Yury: 106
Glenn, John: *18,* 20, *20,* 23
GMS (Japan): 106, 205
Gorbatko, Victor: 53, 106, 130, 135
Gordon, Richard: 38, 53
Grabe, Ronald: 189
Grechko, Georgi: 93, 106, 114, 182, 189
Gregory, Frederick: 189
Great Red Spot: 124, *128, 129*
Grissom, Virgil 'Gus' : *18, 18*, 32, 37, 42, *42*
Gubarov, Alexei: 93, 114, 119
Gurragcha, Jugderdemidyin: 140, 143
Haise, Fred: 56, 106
Hart, Terry: 177
Hartsfield, Henry Jr: 155, *171,* 177, 189

Hauck, Frederick: 165, 177
Hawle, Steven: 177
Hawley, Steven: *171,* 195
Henize, Karl: 189
Hermaszewski, Miroslaw: 114, *116,*119
Hilmers, David: 189
Hoffman, Jeffry: 189
van Hoften, James: 170, 177, 189
Insat (India): 204
Intasat: *79*
Intelsat Series: 32, *34, 35, 37, 66, 67, 90, 91*
Io: 124, *128, 129*
IRAS: 165, *166-167*
Irwin, James: 58
ITOS-G (USA): 78
Ivanchenkov, Alexander: 114, *115,* 119, 155
Ivanov, Georgi: 124, *124,* 127
Jaehn, Sigmund: 114, *116,* 119
Jarvis, Gregory: *192,* 195
Jupiter C: 10
Kerwin, Joseph: 75
Keyhole (US Military): 205
Khrunov, Yevgey: 50, 53
Kizim, Leonid: 75, 130, 135, 170, 177, 195,197
Klimuk, Pyotor: 93, 114, *116,* 119
Komarov, Vladimir: *28,* 31, 42, 44
Kovalenok, Vladimir: 106, 114, *115,* 119, 140, 143
Kubasov, Valeri: 53, 84, 93, 130, 135, *135,* 140
Kvant (USSR): 200
LAGEOS: *98,* 99
Landsat (NASA): 205
Landsat: *91,* 92, 92, 93, *114,* 119, 200
Laveikin, Alexander: 200
Lazarev, Vasily: 75, 93
Leasat (US Military): *173,* 177, 189, 205
Lebedev, Valentin: 75, 155
Leetsma, David: 177
Lenoir, William: *152-153,* 155
Leonov, Alexei: 32, 37, 84, *84, 87,* 93
Lichtenberg, Byron: 165
Lind, Don: 189
Long March: 200
Lounge, Michael: 189
Lousma, Jack: *73,* 75, 155
Lovell, James: 32, 37, 38, 46, 49, 56
Lucid, Shannon: 189
Luna 1: 12, *12, 13*
Luna 2: 12
Luna 3: 12, *12, 13,* 96
Luna 9: 38
Luna 10: 38
Luna 15: 54
Luna 16: 56, *56, 57,* 64
Luna 17: 56
Luna 20: 64, 67
Luna 21: 72, 75
Luna 24: 96
Lunar Orbiters: 38, *44,* 44
Lunar Rovers:69
Lunar Roving Vehicle: 58, 64
Lunokhod 1: *56,* 72
Lunokhod 2: 72, *72*
Lyakhov, Vladimir: 124, 127, 158, 165
Makarov, Oleg: 75, 93, 119, 130, 135, 140
Malyshev, Yuri: 135, 177
Mariner 1: 20
Mariner 2: 20
Mariner 4: 28, *29,* 31, 32
Mariner 10: 75, 78, 81, *82–83*
Marisat (United States): 205
Marisat A: 99, *99, 100, 101*
Marisat B: 100 *101*
Marisat C: 100, *101*
Mars 1: 20
Mars 2: 58
Mars 3: 58, *58, 61, 60–61*
Mars 4: 72, *77*
Mars 5: 72, *77*
Mars 6: *77*
Mars 7: *77*
Mattingly, Thomas: 67, 155
Mendez, Arnaldo Tamayo: 130, 135
Merbold, Ulf: 158, *161,* 165, 190
Mercury: 12, *15*
Mercury/Little Joe 3: 12
Mercury 3/Freedom 7: 18, *18*
Mercury 4: 18
Mercury 5: 18
Mercury 6: 22, 23
Mercury 7: 23, 24
Mercury 8: 23, 24
Mercury 9: 27
Messerschmid, Ernst: 184, 189

Meteor 2: 197, 205
Meteosat (European Space Agency): 204
Midas 2: 14
Mir Space Station: 195, *196*, 197, 200, *200*
Mitchell, Edgar: 58
Morelos (Mexican): *188* 205
Mullane, Richard: *171*, 177
Musgrave, Story: 165, 189
McAuliffe, Sharon Christa: 192, *192*, *193*, 195
McBride, Jon: 177
McCandless, Bruce: *173*, *176*, 177
McDivitt, James: 32, *32*, 37, 50, 53
McNair, Ronald: 177, *192*, 195
Nagel, Steven: 189
Navstar (US Military): 205
Nelson, Bill: 195
Nelson, George: 170, 177, 195
Neri, Rodolfo: 189
Nikolayev, Andrian: 20, 23, 56
Nimbus: 93, *93*
NOAA (NASA): 205
Noctis Labyrinthus: *104*
Ockels, Wubbo: *184*, 189
O'Connor, Bryan: 189
Onizuka, Ellison: 189, *192*, 195
OSO Spacecraft: 93, *95*
Osumi: 56
Overmyer, Robert: *152–153*, 189
Palapa (Indonesia): 106, *110*, *163*, 165, 169, *179*, 204
Pailes, William: 189
Parker, Robert: 165
Patsayev, Viktor: 58
Payton, Gary: 189
Pavonis Mons: *104*
Peterson, Donald: 165
Pioneer 1: 10
Pioneer 10/11: 205
Pioneer 10: 64, 67, *77*, 96
Pioneer 11: *1*, 75, 77, *77*
Pioneer Venus 1: 114, 119, *119*
Pioneer Venus 2: 114, *118*, 119
Planet-A (Japan): *187*, 189, 205
Pogue, William: 75, 77, 78
Popov, Leonid: 130, *132*, 135, 140, 143, 155
Popovich, Pavel: 20, 23, 78
Progress Spacecraft (USSR): 114, 130, 205
Prospero (United Kingdom): 58, 61
Proton (Soviet): 200
Prunariu, Dmitri: 140, 143
Ranger Moon Probes: *23*, 28, *30*, 31, 32
Redstone: 8, 10
Relay 1: *23*, 23
Remek, Vladimir: 114, *116*, 119
Resnik, Judith: 170, *171*, 177, *192*, 195
Rhea: *148–149*
Ride, Sally: 158, *158*, 165. 170, 177, 190
Romanenko, Yuri: 106, 114, 130, 135, 140, 200
Ross, Jerry: 189
Rozhdestvensky, Valery: 99
Rukavishnikov, Nikolai: 78, 124, 127
Ryumin, Valery: 106, 124, 127, 130, *132*, 135, 140, 197
Salyut Space Stations (early): 58, 61, 78, *78*, *79*, 84, 96, 99, 106
Salyut 6: 106, *106*, 114, 115, 124, 130, *130–131*, 132, 140
Salyut 7: 150, *155*, 158, 170, 182, 184, 197
Sarafanov, Gennady: 78
SATCOM Spacecraft: *96*, 99, *188*, 189
Satellite Business Systems (SBS Spacecraft) (United States): *133*, *134*, 135, 150, 153, 170, *174*, *175*, 177, 205
SBTS (Brazil): 204
Saturn: *136-137*, 140, *148*, *149*
Saturn 1B: *69*, 75
Saturn 5: *58*, *69*
Savinykh, Victor: 140, 143, *184-185*, 189
Savitskaya, Svetlana: 150, 155, 170, 177
Schirra, Walter (Wally): *18*, 20, 23, 32, 37, 46, 49, *49*
Schmitt, Harrison: 64, 67
Schweickart, Russell: 50, 53
Scobee, Francis: 177, *192*, 195
Scott, David: 38, 53, 58
Scully-Power, Paul: 177
Serebrov, Alexander: 155, 165
Sevastyanov, Vitali: 56, 93
Seddon, Rhea: 189
Sharma, Rakesh: 170, 177, 200
Shatalov, Vladimir: 50, 53
Shaw, Brewster Jr: 165, 189
Shepard, Alan: 18, *18*, 58
Shonin, Georgi: 53
Shriver, Loren: 189
Skylab: 32, *72*, 74, 75, *75*, 77, 114
Skynet Series (United Kingdom): 205
Slayton, Donald "Deke": *18*, 24, 84, *84*, *87*, 93
Smith, Michael: *192*, 195

Solar Maximum Mission (SMM): *132*, 135, *135*, 170
Solovyev, Vladimir: 170, 177, 195, 197
Soyuz (USSR): 42, 44, 192
Soyuz 2: 46
Soyuz 3: 46, 49
Soyuz 4: 50, 53, *53*
Soyuz 5: 50, 53, *53*
Soyuz 6: 53, *53*, 54, 84
Soyuz 7: 53, 54, 158
Soyuz 8: 53, 54
Soyuz 9: 56
Soyuz 10: 58, 61
Soyuz 11: 58, 61, 77, 130, 192
Soyuz 12: 75, 77
Soyuz 13: 75, 77
Soyuz 14: 78
Soyuz 15: 78
Soyuz 16: 78
Soyuz 17: 84, 93
Soyuz 18: 84
Soyuz 18B: 84, 93
Soyuz 19/ASTP: 84, *86*, *87* 93 *93*
Soyuz 20: 84
Soyuz 21: 96, 99
Soyuz 22: 96, 99
Soyuz 23: 96, 99
Soyuz 24: 106
Soyuz 25: 106
Soyuz 26: 106, 114
Soyuz 27: 114, 119
Soyuz 28: 114, 119
Soyuz 29: 114, 115, 119
Soyuz 30: 114, 119
Soyuz 31: 114, 116, 119
Soyuz 32: 124, 127
Soyuz 33: 124, 127
Soyuz 34: 124
Soyuz 35: 130, 131, 135
Soyuz 36: 130, 131, 135
Soyuz 37: 130, 131, 135
Soyuz 38: 130, 131, 135
Soyuz 39: 140, 143
Soyuz 40: 140, 143
Soyuz T (USSR): 205
Soyuz T-1: 124
Soyuz T-2: 130, 135
Soyuz T-3: 130
Soyuz T-4: 140, 143
Soyuz T-5: 150, 155
Soyuz T-6: 150, 155
Soyuz T-7: 150, 155
Soyuz T-8: 158, 165
Soyuz T-9: 158, 165
Soyuz T-10: 158, 170
Soyuz T-10B: 170, 177
Soyuz T-11: 170, 177
Soyuz T-12: 170, 177
Soyuz T-13: 182, 184, 189
Soyuz T-14: 182, 184, 189
Soyuz T-15: 195, 197
Soyuz TM-1 (USSR): 197, 205
Soyuz TM-2: 200
Soyuz TM-3: 200
Spacenet (United States): 205
Space Shuttle (NASA): 32, 124, *124*, 130, 140, 158, 174, 175, 192, 197, 200, 205, See also STS *Atlantis*, *Challenger*, *Columbia*, *Discovery* and *Enterprise*
SPAS (Federal Republic of Germany): 158, 169, 204
SPOT (France): 195, 204
Spring, Sherwood: 189
Sputnik 1: 8, *8*
Sputnik 2: 8
Sputnik 3: 10, *10*
Sputnik 5: 14, 15
Sputnik 6: 14
Stafford, Tom: 32, 37, 38, *39*, 50, 53, 84, *84*, 93
Stewart, Robert: 177, 189
STS-1: *141*, 142, 143, 145, *145*
STS-2: 140, 142, 143, 145
STS-3: 150, 155
STS-4: 150, 155, 158
STS-5: 150, 151, *153*, 155
STS-6: 158, 165
STS-7: 158, *159*, *160*, 163, 165, 169
STS-8: 158, *161*, *162*, *163*, 165
STS-9: 158, 165
STS-41-B: 170, 176, 177
STS-41-C: 170, 177
STS-41-D: 170, 173, 177
STS-41-E: 170
STS-41-F: 170
STS-41-G: 170, 177
STS-51-A: 170, 177, 179

STS-51-B: 182, 189
STS-51-C: 189
STS-51-D: 189
STS-51-F: 189
STS-51-G: 182, *182*, 184, 189
STS-51-I: 189
STS-51-J: 189
STS-51-L: 192, *192*
STS-61-A: 182, 184, 189
STS-61-B: *188*, 189
Strekalov, Gennady: 130, 135, 158, 165, 177
Sullivan, Kathryn: 170, 177
Surveyor spacecraft: 38, *38*, 42, *43*, *44*, 44
Swigert, John: 56
Syncom 1: 24, *26*, 27
Syncom 2: 24, *26*
Syncom 3: 28, 31
Syncom 4 (Leasat 2): 170, *172*
Telecom (France): 204
Tele-X (Sweden): 205
Telstar (United States): 205
Telstar 1: 20, *21*, 23
Telstar 2: 24
Telstar 3: *164*, 170
Telstar 3A: 165
Tereshkova, Valentina: 24, *24*, 27, 150
Tethys: *136*, *148-149*
Thagard, Norman: 165, 189
Tharsis Ridge: *102*, *104*
Thornton, William 165, 189
TIROS 1: 14, *14*, 15
TIROS 8: 24, *25*
TIROS-N: *114*, 119
Titan: *136*, 197
Titov, Vladimir: 158, 165
TOS/ESSA: *48*, *49*
Transit spacecraft: 14
Truly, Dick: 106, 140, *142*, 143, *145*, 165
Tuan, Pham: 135
Ulysses Patera: *104*
Van Den Berg, Lodewijk: 189
Vanguard: 8, *9*, 10, *10*
Vasyutin, Vladimir: *184-185*, 189
Vega: 177, *179*, 197
Venera 4: *44*, 44
Venera 9: 93, *93*
Venera 10: *93*
Venera 13/14: *156*
Viking (Sweden): 205
Viking 1: 84, *88*, 93, 96, *103*
Viking 2: 84, *88*, *89*, 93, 96, *102–103*
Viktorenko, Alexander: 200
Volkov, Alexander: *184–185*, 189
Volkov, Vladislav: 53, 58
Volynov, Boris: 50, 53, 99
Voskhod 1: 28, 31
Voskhod 2: 32, 37, 84
Vostok 1: 18
Vostok 2: 18
Vostok 3: 23
Vostok 4: 23
Vostok 5: 24, 27
Vostok 6: 27
Voyager 1: 106, 124, 130, 136
Voyager 2: 106, 124, 140, 148, 149, 192
Voyager spacecraft (general references): 4, *107*, *108–109*, *111*, 130, 205
Walker, Charles: *171*, 177, 189
Walker, David: 177
Wang, Taylor: 189
Weitz, Paul: 75, 165
Westar (United States): 205
Westar 1: 78, *79*
Westar 2: 80
Westar 3: 80, 127, *127*
Westar 6: *178–179*
White, Ed: 32, *32*, 37, 42, *42*
Williams, Donald: 189
Wresat (Australia): 42, 44
Yegorov, Boris: 28, 31
Yeliseyev, Aleksei: 50, 53, 67
Young, John: 32, 37, 38, 50, 53, *62–63*, 64, 140, *142-143*, 143, *145*, *161*, 165, 190
Zholobov, Vitaly: 99
Zond (spacecraft): 46, *53*
Zudov, Vyacheslav: 99

Plans for the future. *Overleaf:* **This Rockwell International conceptual illustration shows a dual keel 'Power Tower' space station, which would essentially be an orbital high voltage power station for use by various orbital NASA systems. Note the Shuttle Orbiter, which would routinely deliver both supplies and personnel.**